Technikzukünfte, Wissenschaft und Gesellschaft / Futures of Technology, Science and Society

Edited by
A. Grunwald, Karlsruhe, Germany
R. Heil, Karlsruhe, Germany
C. Coenen, Heidelberg, Germany

Diese interdisziplinäre Buchreihe ist Technikzukünften in ihren wissenschaftlichen und gesellschaftlichen Kontexten gewidmet. Der Plural „Zukünfte" ist dabei Programm. Denn erstens wird ein breites Spektrum wissenschaftlich-technischer Entwicklungen beleuchtet, und zweitens sind Debatten zu Technowissenschaften wie u.a. den Bio-, Informations-, Nano- und Neurotechnologien oder der Robotik durch eine Vielzahl von Perspektiven und Interessen bestimmt. Diese Zukünfte beeinflussen einerseits den Verlauf des Fortschritts, seine Ergebnisse und Folgen, z.B. durch Ausgestaltung der wissenschaftlichen Agenda. Andererseits sind wissenschaftlich-technische Neuerungen Anlass, neue Zukünfte mit anderen gesellschaftlichen Implikationen auszudenken. Diese Wechselseitigkeit reflektierend, befasst sich die Reihe vorrangig mit der sozialen und kulturellen Prägung von Naturwissenschaft und Technik, der verantwortlichen Gestaltung ihrer Ergebnisse in der Gesellschaft sowie mit den Auswirkungen auf unsere Bilder vom Menschen.

This interdisciplinary series of books is devoted to technology futures in their scientific and societal contexts. The use of the plural "futures" is by no means accidental: firstly, light is to be shed on a broad spectrum of developments in science and technology; secondly, debates on technoscientific fields such as biotechnology, information technology, nanotechnology, neurotechnology and robotics are influenced by a multitude of viewpoints and interests. On the one hand, these futures have an impact on the way advances are made, as well as on their results and consequences, for example by shaping the scientific agenda. On the other hand, scientific and technological innovations offer an opportunity to conceive of new futures with different implications for society. Reflecting this reciprocity, the series concentrates primarily on the way in which science and technology are influenced social and culturally, on how their results can be shaped in a responsible manner in society, and on the way they affect our images of humankind.

Prof. Dr. Armin Grunwald, Physiker, Mathematiker und Philosoph, lehrt Technikphilosophie und Technikethik am Karlsruher Institut für Technologie (KIT), ist Leiter des Instituts für Technikfolgenabschätzung und Systemanalyse (ITAS) in Karlsruhe und Leiter des Büros für Technikfolgen-Abschätzung beim Deutschen Bundestag (TAB) in Berlin. / Professor Armin Grunwald, physicist, mathematician and philosopher, teaches the philosophy and ethics of technology at the Karlsruhe Institute of Technology (KIT), and is the director of the Institute for Technology Assessment and Systems Analysis (ITAS) in Karlsruhe and of the Office of Technology Assessment at the German Bundestag (TAB) in Berlin.

Reinhard Heil, Philosoph, ist wissenschaftlicher Mitarbeiter am KIT-ITAS. / Reinhard Heil, philosopher, is a researcher at KIT-ITAS.

Christopher Coenen, Politikwissenschaftler, ist wissenschaftlicher Mitarbeiter am KIT-ITAS und Herausgeber der Zeitschrift 'NanoEthics: Studies of New and Emerging Technologies'. / Christopher Coenen, political scientist, is a researcher at KIT-ITAS and the editor-in-chief of the journal 'NanoEthics: Studies of New and Emerging Technologies'.

Joachim Boldt (Ed.)

Synthetic Biology

Metaphors, Worldviews,
Ethics, and Law

Editor
Joachim Boldt
Albert-Ludwigs-Universität Freiburg
Freiburg, Germany

Technikzukünfte, Wissenschaft und Gesellschaft / Futures of Technology, Science and Society
ISBN 978-3-658-10987-5   ISBN 978-3-658-10988-2 (eBook)
DOI 10.1007/978-3-658-10988-2

Library of Congress Control Number: 2015954947

Springer VS
© Springer Fachmedien Wiesbaden 2016
This work is subject to copyright. All rights are reserved by the Publisher, whether the whole or part of the material is concerned, specifically the rights of translation, reprinting, reuse of illustrations, recitation, broadcasting, reproduction on microfilms or in any other physical way, and transmission or information storage and retrieval, electronic adaptation, computer software, or by similar or dissimilar methodology now known or hereafter developed.
The use of general descriptive names, registered names, trademarks, service marks, etc. in this publication does not imply, even in the absence of a specific statement, that such names are exempt from the relevant protective laws and regulations and therefore free for general use.
The publisher, the authors and the editors are safe to assume that the advice and information in this book are believed to be true and accurate at the date of publication. Neither the publisher nor the authors or the editors give a warranty, express or implied, with respect to the material contained herein or for any errors or omissions that may have been made.

Lektorat: Frank Schindler, Monika Mülhausen

Printed on acid-free paper

Springer VS is a brand of Springer Fachmedien Wiesbaden
Springer Fachmedien Wiesbaden is part of Springer Science+Business Media
(www.springer.com)

Preface

The results, arguments, and theses that are presented in this volume have been developed and discussed within the project *Engineering Life: An interdisciplinary approach to the ethics of synthetic biology*. The project was funded by the German Ministry of Research and Education (grant nr. 01GP1003). The following contributors were part of this project. In the order of appearance: Joachim Boldt, Oliver Müller, Harald Matern, Jens Ried, Matthias Braun, Peter Dabrock, Tobias Eichinger, Jürgen Robienski, Jürgen Simon, Rainer Paslack, Harald König, Daniel Frank, Reinhard Heil, Christopher Coenen.

Jan C. Schmidt, Bernadette Bensaude Vincent, Johannes Achatz, Iñigo de Miguel Beriain, Sacha Loeve, Christoph Then, Bernd Giese, Henning Wigger, Christian Pade, Arnim von Gleich joined us at different occasions, first and foremost our concluding conference at Freiburg University, Germany.

I'd like to thank the all the project partners and contributors for their cooperation. I am grateful to the German Ministry of Research and Education for giving us all the opportunity to engage in philosophical, ethical, legal, and social science research relating to synthetic biology. I'd also like to thank the Centre for Biological Signalling Studies (BIOSS) at Freiburg University for their kind support.

Special thanks to Susan Keller for skillfully adjusting the English language style, grammar and spelling in this volume, and to Sebastian Höpfl for accurately unifying bibliography styles, citations, and layout.

Freiburg, August 28, 2015   Joachim Boldt

Contents

Swiss watches, genetic machines, and ethics. An introduction to synthetic
biology's conceptual and ethical challenges 1
Joachim Boldt

I Concepts, Metaphors, Worldviews

Philosophy of Late-Modern Technology. Towards a Clarification and
Classification of Synthetic Biology 13
Jan C. Schmidt

Synthetic Biology. On epistemological black boxes, human self-assurance,
and the hybridity of practices and values 31
Oliver Müller

Living Machines. On the Genesis and Systematic Implications of
a Leading Metaphor of Synthetic Biology 47
Harald Matern, Jens Ried, Matthias Braun and Peter Dabrock

Production biology. Elements and limits of an action paradigm
in synthetic biology 61
Tobias Eichinger

Creativity and technology. Humans as co-creators 71
Harald Matern

The moral economy of synthetic biology 87
Bernadette Bensaude Vincent

Evaluating biological artifacts. Synthetic cells in the philosophy
of technology 101
Johannes Achatz

II Public Good and Private Ownership. Social and Legal Ramifications

Legal Aspects of Synthetic Biology 123
Jürgen Robienski, Jürgen Simon and Rainer Paslack

Synbio and IP rights: looking for an adequate balance between
private ownership and public interest 141
Iñigo de Miguel Beriain

III Opportunities, Risks, Governance

Beyond unity: Nurturing diversity in synthetic biology and its publics 155
Sacha Loeve

Synthetic Genome Technologies 185
Christoph Then

Promising applications of synthetic biology – and how to avoid their
potential pitfalls 195
Bernd Giese, Henning Wigger, Christian Pade and Arnim von Gleich

Synthetic biology's multiple dimensions of benefits and risks:
implications for governance and policies 217
Harald König, Daniel Frank, Reinhard Heil and Christopher Coenen

Contributors 233

Swiss watches, genetic machines, and ethics
An introduction to synthetic biology's conceptual and ethical challenges

Joachim Boldt

1 Novelty and conceptual framing of emerging technologies

Emerging technologies have a history. They do not emerge out of nothing but develop gradually and continuously. Synthetic biology is no exception. It is rooted in genetic engineering, and many observers maintain that synthetic biology is no more than a new label for just that: genetic engineering.

The inevitable question therefore arises: when is an emerging technology in fact a new technology and when is it only a gradual development of an already known technology? Part of the answer certainly does not lie in the technology itself but in the context of interests surrounding it. A supposedly new technology comes with new economic and societal opportunities – and new risks. That is to say, a new label suits researchers who are seeking to obtain grants just fine. At the same time, it suits technology critics, too.

Again, synthetic biology is no exception to this rule. Scientists use the label "synthetic biology" to set it apart from so-called traditional genetic engineering, which implies that synthetic biology is a non-traditional, modern, and particularly capable technology (Knight 2005). NGOs such as the ETC Group, on the other hand, equate the new label with "extreme genetic engineering", a description in which the adjective "extreme" leads to associations of highly risky undertakings (ETC 2007).

True as this may be, it nonetheless overstates the case if one explains the formation of a new emerging technology label entirely in terms of accompanying interests. Even if the technology itself is in a process of gradual development, the ways in which these technologies are understood and thus further developed, and the concepts in the light of which technologies are formed, may shift in leaps and bounds. Human beings are truth seeking animals, and their understanding of what concepts come closest to a true description of their study objects and intentions has an impact on the further development of a technology, just as economic or other interests that are only arbitrarily connected to a science and technology do.

Mainstream synthetic biology incorporates such a shift in conceptual framing. It aspires to move away from genetic engineering guided by trial and error towards a rational design process in which whole genomes can be constructed at the computer. Fast and cheap whole genome sequencing, reliable and affordable gene synthesis, and highly effective genome editing methods such as CRISPR/Cas9 pave the way towards realizing these objectives.

To rationally and reliably design and assemble a complex entity presupposes that the behavior and functions of the complex entity can be predicted on the basis of the behavior of its parts. Many artificial technological objects adhere to this requirement, as we know from experience. With regard to living beings, though, one may suppose that the case is different. After all, living beings are subject to evolutionary change; they interact in multiple ways with their environment; and we even attribute freedom of will to some of them, namely ourselves. Thus, to suppose that living beings can be reliably designed amounts to an ontological hypothesis claiming that the behavior of living beings can be explained in terms of the behavior of their physical parts.

By itself, this conceptual frame is far from new. It can be traced back at least to 17th century Cartesian philosophy and the mechanical animal models of the Renaissance. However, it was not until the discovery of DNA in the 20th century and today's advanced genome editing abilities that purposeful building and rebuilding of long DNA sequences and whole genomes became feasible (Keller 2002). Today, these assumptions form an epistemic, i.e., truth related, conceptual frame that is shaping synthetic biology and guiding it into the future. This frame differs from the guiding assumptions and goals of traditional genetic engineering, which are less systematic and less engineering and design oriented. It is precisely the specific conceptual frame of synthetic biology that adds a decisive element towards an exhaustive explanation of why synthetic biology bears a label of its own.

Building and rebuilding DNA differs from building mechanical animals. A mechanical animal may be regarded as being analogous to, but certainly not identical to, a natural animal, since its physical parts differ from the parts of a natural organism. Building the synthetic genome of a single-cell organism, by contrast, amounts to assembling parts that make up natural organisms as well. Again, in contrast to mechanical animals, the prospect of organisms created by synthetic biology urges the question of whether such a synthetic living being that is indiscernible from a natural living being with regard to its physical and, as one must assume, its emergent properties can be adequately understood as a quasi-mechanical object.

Organisms of synthetic biology inhabit a curious space between artificial objects and natural living beings because they are living beings, albeit constructed and treated as if they were not. If it is adequate to conceptualize living beings as quasi-mechanical objects, then synthetic biology is the technology that ultimately unsettles our – supposedly wrong, from synthetic biology's point of view – deeply entrenched everyday understanding that living beings are something different from, and more than, non-living entities. In addition, the synthetic biology conceptual frame has a bearing on the way in which one perceives ethical and societal impacts of this technology. It is an important task of bioethics, the social sciences, philosophy, and theology to identify, point out, and reckon with these impacts, especially in the early stages of a technological development.[1]

2 Synthetic biology, nanotechnology, and a conceptual challenge

In its quest to reveal the hidden secrets of an object in the object's basic parts, synthetic biology resembles nanotechnology. Nanotechnology, too, is animated by the idea of being able to build and rebuild complex objects by rearranging their parts. The title of the U.S. National Science and Technology Council's (NSTC) report "Nanotechnology. Shaping the world atom by atom" bears witness to this parallel. Unlike synthetic biology, though, nanotechnology literally aims to engineer machines. Nanoscale engineering objects are meant to be non-living entities that are put together at an atomic level. The NSTC does not go to great effort to assess the scientific feasibility of this ideal, but straightforwardly compares nanoscale objects to "Swiss watches": "The products of Swiss watchmakers even several centuries ago proved that human control over the material world had extended downward a thousandfold to the millimeter scale or so. Over the past few decades, researchers have pushed this control down another hundredfold" (NSTC 1999, p.5).

Synthetic organisms, by contrast, are by definition living beings whose nanoscale DNA components have been rearranged in novel ways. One may think of a synthetic cell as the "best shot at a general nanotech assembler, the dream of Eric Drexler and many nanotechnology enthusiasts" (Church and Regis 2012, pp.53f.), but the synthetic cell itself does not constitute a nanotechnology product as it is commonly defined. Even if one envisages the assembling of a functional cell from

---

1 The main focus of parts I and II of this volume is directed towards such normatively relevant conceptual issues within synthetic biology.

non-living molecules, this is a synthetic biology task, not a nanotechnological one. Nanobiotechnology, the area within nanotechnology closest to biology, comprises the engineering of nanoscale machines, i.e., non-living objects, that make use of or are inspired by biological molecules.

From an ethical perspective, nanotechnology and nanobiotechnology thus present issues of technological control and risk assessment. When confronted with nanotechnology one has to ask: can we responsibly do what we aim to do? Synthetic biology leads to the further question: do we know what we are talking about when we conceptually align the living world with the non-living world? Synthetic biology – bionanotechnology, if you like – is a challenging enterprise not only with regard to risk assessment but also with regard to the use of concepts and metaphors.

If one distinguishes nanotechnology and synthetic biology in this way, the distinction hinges upon the difference between non-living and living entities being as clear-cut as possible. Notwithstanding the everyday self-evidence of this distinction, attempts to spell out precisely what sets the living world apart from the world of non-living entities have kept philosophers and scientists busy since the ancient beginnings of philosophical and scientific thinking.

Biology textbooks typically offer open lists of criteria for what counts as life. These include, for instance, metabolism, homeostasis, growth, stimulus response, reproduction, and adaptation and evolution. When one looks for necessary and adequate properties to explain and sort these criteria one often comes across concepts such as "self-organization" or "self-preservation" (Bensaude Vincent 2009). The prefix "self" is important here because it carries the idea that the behaviors of living objects cannot be explained purely with reference to determining causal forces but need to include the notion of something *reacting to* something else outside of itself. Living beings require a shift from causal explanations towards telos-oriented ones, it is assumed (Boldt 2013).

Which kind of abilities justify such a shift to a telos-oriented explanatory scheme, and whether we are able to reliably detect their presence, remains a debatable issue. Nonetheless, if one maintains that there is indeed a difference between living and non-living entities, the difference will have to be spelled out in terms of these notions.

Heuristically at least, synthetic biology itself maintains the distinction. It is a biotechnology explicitly aiming at restructuring and reinventing the molecular basis of *life*. It does so because objects that possess metabolism, reproduce, and undergo evolutionary change can be more efficient and more powerful technological tools than non-living technologies. At the same time, these properties come with a price. Besides the fact that growth is energy-consuming and reproduction can be unreliable, evolution implies a certain degree of independence of the engineered object's behavior and development from the engineer's initial plans and intentions, to name one ethically relevant property of living beings.

Applying the synthetic biology conceptual frame may lead to an underrating of the relevance of the above point. As long as synthetic biology has not advanced to a stage at which it becomes possible to evaluate single applications, it is one of the most important tasks of bioethics to analyze the conceptual framework of this technology, describe its limits, and compare it to alternative accounts. The engineering and machine framework of synthetic biology and the faith that synthetic biologists place in it certainly deserve such scrutiny, as indicated by statements such as the following:

> Originally, you pretty much had to take organisms as they came, with all the inherent design flaws and limitations, compromises and complications that resulted from the random working of evolution. Now we could actually preplan living systems, design them, construct them according to our wishes, and expect them to operate as intended – just as if they were in fact machines (Church and Regis 2012, p.182).

As becomes evident, the machine metaphor and the engineering framework of synthetic biology shape one's perspective on the abilities of synthetic biology and on the function and behavior of its objects. DNA segments that appear redundant and DNA expression pathways that seem to be unnecessarily complicated are not deemed to call for closer analysis but are classified as flaws. In the same vein, the future behavior and development of a synthetic organism is thought to be safely following the engineer's intentions, disregarding the possibility of unexpected evolutionary change. The synthetic biologists de Lorenzo and Danchin, who otherwise share an optimistic outlook for synthetic biology's general and long-term ability to design reliable synthetic organisms with predictable properties and behaviors, object to the mainstream conceptual frame of their research field: engineering metaphors for gene expression, for example, "represent a straight and overtly simplistic projection of electric engineering concepts into supposedly biological counterparts," they write (Lorenzo and Danchin 2008, p.824).

3 Synthetic biology and existing GMO regulation

Before turning to the possible ethical and societal impact of the current synbio conceptual frame in more detail, it is worth noting that the emergence of synthetic biology takes place within a set of existing laws and regulations, nationally and internationally. Exaggerated promises and expectations notwithstanding, synthetic biology may indeed soon offer useful applications. There are promising approaches within the field of medicine, and energy and the environment are further fields of application.[2] Each application will of course have to conform to established legal and ethical regulation. With regard to short-term applications of synthetic biology, the relevant fields are covered to a large extent by a number of existing national or international regulations.[3]

In the long run, however, synthetic biology's research agenda may lead to products that fall outside the field of established risk assessment procedures and current regulation. Synthetic organisms whose genomes stem from a large number of different sources, for example, aggravate the task of risk assessment. Established risk assessment procedures for genetically modified organisms rely on an evaluation of the known behavior and risk profile of the natural counterpart organism. If the genome of the synthetic organism no longer resembles the genome of any natural species, the basis for a risk assessment procedure of this kind is lacking. What is more, if synthetic biologists one day engineer a synthetic cell containing only non-natural DNA molecules, it will be difficult to classify such an organism as a "genetically modified" one. Most probably, any such organism will not fall within the scope of current GMO regulations (Pauwels et al. 2012).

In all cases it holds that the more encompassing genetic modification and replacement become, the more difficult it will be to assess the risks and apply existing regulations. One way out of the risk assessment challenge may be to encourage step by step genome changes in order to ensure that every novel synthetic organism does have a like and known predecessor. With regard to non-natural DNA organisms, modification of legal regulations will be unavoidable.

4   The synbio story, ethics, and the diversity of research approaches

Assessing applications according to legal and ethical standards is an important ethical, societal, and political task, but not the only one. Given synthetic biology's powerful self-narrative that is shaping the future of this emerging technology as well as its perception of its objects – it is equally important to be attentive to this

---

2   Societally relevant synbio application scenarios are described by König (in this volume).
3   An overview of legal regulations is supplied by Robienski, Simon, and Paslack (in this volume). Cf. also Kuzma and Tanji (2010).

very narrative, its limits, and alternative accounts of what synthetic biology may be and become.

The constraints of the machine metaphor and the engineering framework of synthetic biology are ethically relevant. For example, how one rates the threat to an existing ecosystem posed by a synthetic organism depends on, among other assumptions, how accurately one believes oneself able to predict the development of the synthetic species. To name another example, thinking along the lines of the design and engineering approach does not restrict synthetic biology to engineering single-cell organisms. On the contrary, it appears natural from this perspective to expect synthetic biology to extend its scope to higher organisms, including humans, as soon as this appears technically feasible. From the contested biocentrist's point of view, ethical questions regarding inherent value and instrumentalization are relevant – to a higher or lesser degree – whenever living beings are subjected to technological interventions according to ends that do not conform to the ends and interests of the organism itself (Deplazes-Zemp 2012). Nonetheless, when one considers human beings, issues of instrumentalization inevitably become relevant, regardless of whether one is operating from a biocentric or anthropocentric ethical foundation. From an organism-as-machine standpoint, these issues are difficult to recognize and understand, let alone tackle.

Paying attention to the current synbio self-narrative, its limits, and alternative accounts of what synthetic biology may be, could contribute to developing accounts of synthetic biology that put less emphasis on the strong design and engineering framework found today. Synthetic biology need not be understood as a technology aiming to rationally design a second nature (Keller 2009). It can also be framed as a technology inspired by and mimicking natural organisms and natural processes of DNA change. Such an interpretation and its application ideals may, for example, help to alleviate safety concerns. One may also envisage synthetic biology as a technology that aims to use DNA and its products without relying on those energy consuming and difficult to control processes such as evolution and growth that are inherently bound to the phenomenon of life (Giese and Gleich 2014). In this case, cell-free systems would count as synthetic biology's poster child. Again, this approach may be seen as safer when compared to synthetic organisms. Cell-free approaches are obviously also less likely to fall victim to concerns that, once single-cell organisms have been equipped with novel synthetic genomes, animals and humans come next.

What the search for alternative stories to describe the core aims and sources of synthetic biology amounts to, then, is an appeal for a diversification of research agendas and research approaches.[4] One may dispute any such story and its research approaches, but being able to recognize a variety of stories and approaches is valuable in itself since it opens one's eyes to the advantages and limits of each one.

Developing alternative accounts will also help to put the high hopes and promises accompanying the advent of synthetic biology into perspective. If synthetic biology is not (or at least not necessarily and exclusively) about freeing oneself from the limits of natural organisms and DNA in order to engineer life that perfectly matches one's wishes, it may instead be seen as one technological tool among other technological tools and social measures, each of which may contribute its share to overcoming societal challenges. Public debate will profit from this view just as much as synthetic biology itself.

5 A closing remark

When we, literally or conceptually, aspire to turn living nature into our tool, we ultimately turn our own origin into a tool. The inconsistency of this project comes to the fore most clearly when we direct it at our own nature. We are, and must always be, simultaneously the subject and object of our nature. By being unaware of this reality we risk fixating our own development on arbitrary ends. The effect would be that we would become prone to falling victim to those arbitrary ends.

Setting ourselves apart from the world of non-human life is easier. But still, that very world has given birth to us. It contains the seeds of all of our highest human abilities. We do not know what other valuable states it may lead to. If we attempt to fixate nature's ends on our own, we may, to our own disadvantage, miss important developmental properties of living beings and hinder the evolution of many sources of unexpected value. That is not what synthetic biology need or ought to be about.

---

4   An appeal that is convincingly made by Loeve (in this volume).

References

Bensaude Vincent, B. (2009). Self-Assembly, Self-Organization: Nanotechnology and Vitalism. *Nanoethics, 3*(1), 31-42. doi: 10.1007/s11569-009-0056-0.

Boldt, J. (2013). Life as a Technological Product: Philosophical and Ethical Aspects of Synthetic Biology. *Biological Theory, 8*(4), 391-401. doi: 10.1007/s13752-013-0138-7.

Church, G.M., & Regis, E. (2012). *Regenesis. How synthetic biology will reinvent nature and ourselves*. New York: Basic Books.

Deplazes-Zemp, A. (2012). The Moral Impact of Synthesising Living Organisms: Biocentric Views on Synthetic Biology. *Environmental Values, 21*(1), 63-82. doi: 10.3197/09632711 2X13225063228023.

ETC (2007). ETC Group. *Extreme Genetic Engineering. An Introduction to Synthetic Biology*. http://www.etcgroup.org/sites/www.etcgroup.org/files/publication/602/01/synbioreportweb.pdf. Accessed: 26 March 2015.

Giese, B., & Gleich, A. v. (2014). Hazards, risks, and low hazard development paths of synthetic biology. In: B. Giese, C. Pade, H. Wigger, A. von Gleich (eds.), *Synthetic biology. Character and impact* (pp.173-196). Heidelberg: Springer.

Keller, E.F. (2002). *Making sense of life explaining biological development with models, metaphors, and machines*. Cambridge, London: Harvard University Press.

Keller, E.F. (2009). What Does Synthetic Biology Have to Do with Biology? *BioSocieties, 4*(2-3), 291-302.

Knight, T.F. (2005). Engineering novel life. *Molecular Systems Biology, 1*(1). doi: 10.1038/msb4100028.

Kuzma, J., & Tanji, T. (2010). Unpackaging synthetic biology: Identification of oversight policy problems and options. *Regulation & Governance, 4*(1), 92-112. doi: 10.1111/j.1748-5991.2010.01071.x.

Lorenzo, V. de, & Danchin, A. (2008). Synthetic biology: Discovering new worlds and new words. The new and not so new aspects of this emerging research field. *EMBO Reports, 9*(9), 822-827. doi: 10.1038/embor.2008.159.

NSTC (1999). National Science and Technology Council of the USA. *Nanotechnology. Shaping the world atom by atom*. http://www.wtec.org/loyola/nano/IWGN.Public.Brochure. Accessed: 12 May 2015.

Pauwels, K., Willemarck, N., Breyer, D., & Herman, P. (2012). *Synthetic Biology: Latest developments, biosafety considerations and regulatory challenges*. Biosafety and Biotechnology Unit (Belgium). http://www.biosafety.be/PDF/120911_Doc_Synbio_SBB_FINAL.pdf. Accessed: 12 May 2015.

I
Concepts, Metaphors, Worldviews

Philosophy of Late-Modern Technology
Towards a Clarification and Classification of Synthetic Biology

Jan C. Schmidt

1 Two types of technology?

Synthetic biology is the crystallization point of late-modern technoscientific hypes and hopes. In 2010 the research entrepreneur Craig Venter announced the forthcoming advent of an epochal break and envisioned a fundamental shift in our technical capabilities. Synthetic organisms "are going to potentially create a new industrial revolution if we can really get cells to do the production we want; […] they could help wean us off of oil, and reverse some of the damage to the environment like capturing back carbon dioxide" (Venter 2010).[1]

In order to analyze whether the epochal break claims are justified, I will coin a provisional search term and call the (possible) novel kind of technology "late-modern" (Schmidt 2012a; Schmidt 2012b). Apparently, this new type of technology seems to be inherently linked to the concept of self-organization. If such a self-organization based technology is emerging, we have to clarify what is meant by the catchword 'self-organization', and we need to analyze the source or root of self-organization, including the idea(l) of self-productiveness. The thesis is: instabilities — or, in cognate terms, sensitivities — constitute the necessary condition and, hence, the technoscientific core of this type of technology. Based on such analysis, I argue that late-modern technology differs from the classic-modern type of technology in its view and valuation of stability and instability. In fact, this novel kind of technology appears as nature and behaves like nature. In other words, we are experiencing a 'naturalization of technology' in a twofold way, as will be shown in this article. My aim here is to disclose a possible new ambivalence and dialectic of this envisioned late-modern turn in technology for our (late-modern) societies.

---

1 Venter's visionary claim was evidently induced by the success of his team in the *Creation of a Bacterial Cell Controlled by a Chemically Synthesized Genome*—as his article in Science Magazine was titled (Gibson et al. 2010).

2 Clarifying the umbrella term 'synthetic biology'

The exact meaning of the umbrella term 'synthetic biology' is not clear at all. New labels and trendy watchwords generally play a key role in the construction of new technoscientific waves. 'Synthetic biology' is, indeed, an extremely successful buzzword, as was 'nanotechnology' more than one decade ago.[2]

All ethicists and technology assessment scholars are aware of the fact that labels are strongly normative. Labels are not innocent or harmless: they carry content and form the backbones of visions. They are roadmaps towards the future and can quickly turn into reality; they shape the technoscientific field and determine our thinking, perception, and judgment. Labels help to foster hopes and hypes, as well as concerns and fears; their implicit power to create or close new research trajectories and development roadmaps can hardly be overestimated. Labels are part of what could be described as 'term politics' that regulate and shape the field with a 'gate keeper function' to decide who is *in* and who is *out*, in particular, whose research field can be considered as 'synthetic biology' and whose is just part of traditional biotechnology. Labels are relevant with respect to funding, publication opportunities, reputation, and career. Thus, they determine and sway our future, in one way or another. What does the umbrella term 'synthetic biology' mean? Is there a unifying arc and common denominator? What visions do synthetic biologists have, and how likely will their visions be achieved? Three popular definitions of 'synthetic biology', and of what it should be, stand out.

First – goals: The *engineering definition* frames synthetic biology as being radically new since it is said to bring an engineering approach to the scientific discipline of biology. Such an understanding is advocated by a High Level Expert Group of the European Commission: "Synthetic biology is the engineering of biology: the synthesis of complex, biologically based (or inspired) systems […]. This engineering perspective may be applied at all levels of the hierarchy of biological structures […]. In essence, synthetic biology will enable the design of 'biological systems' in a rational and systematic way" (European Commission 2005, p.5). This comes close to the definition given by Pühler et al. who define synthetic biology as "the birth of a new engineering science" (Pühler et al. 2011). Similarly, others view synthetic

---

2   On the one hand, 'synthetic biology' seems to be a fairly young term. It was (re-)introduced and presented by Eric Kool in 2000 at the annual meeting of the American Chemical Society. Since then, the term has gone on to enjoy a remarkable career and general circulation in the scientific communities as well as in science, technology, and innovation politics. On the other hand, the notion of 'synthetic biology' emerged about 100 years ago—although it was rarely mentioned until 2000. It seems more appropriate to consider the more recent understandings of 'synthetic biology'.

biology as "an assembly of different approaches unified by a similar goal, namely the construction of new forms of life" (Deplazes and Huppenbauer 2009, p.58). The engineering definition is generally based on the strong assumption that, before synthetic biology arose, a clear line existed between biology as an academic discipline, on the one hand, and engineering/technical sciences, on the other.[3] The proponents of the engineering definition believe that the well-established divide between the two disciplines is becoming blurred. Today, engineering is transferring its goals to the new subdiscipline of biology. According to the advocates of this definition, these goals have never been characteristics of other subdisciplines of biology. The essential claim is that we are experiencing an epochal break or a qualitative shift within biology: the aim is not theory, but technology.

Second – objects: The *artificiality definition* of synthetic biology is more concerned with objects and material entities than with goals. According to the EU project TESSY, 'synthetic biology' deals with "bio-systems [...] that do not exist as such in nature" (TESSY 2008). In an equivalent sense it is stated that synthetic biology encompasses the synthesis and construction of "systems, which display functions that do not exist in nature" (European Commission 2005, p.5). The German Science Foundation similarly identifies the emergence of "new properties that have never been observed in natural organisms before" (DFG et al. 2009, p.7). "Synthetic biology" is here defined by the non-naturalness, or unnaturalness, and artificiality of the constructed and created bio-objects. *Divergence from nature* appears to be the *differentia specifica* of 'synthetic biology', with 'nature' being seen as the central anchor and negative foil for this definition. Whereas bio-systems were formerly natural, i.e., they occurred exclusively *in*, and were created *by* nature alone, the claim here is that, from now on, they can also be artificial, i.e., constructed by humans.[4]

Third – methods: The *extreme biotechnology definition* leads either to synthetic biology being seen in a more relaxed light or, on the contrary, to it being condemned

---

3   From this angle, biology is regarded as a pure science aiming at fundamental descriptions and explanations. In contrast, engineering sciences appear to be primarily interested in intervention, construction, and creation. Viewed from this angle, biology and engineering sciences have always been—in terms of their goals—like fire and ice.

4   That is certainly a strong presupposition, and it is also linked to the Aristotelian idea of a dichotomy between nature and technical objects. This dichotomy traces back to the Greek philosopher Aristotle who drew a demarcation line between *physis* (nature) and *techné* (arts, technical systems). In spite of Francis Bacon's endeavors at the very beginning of the modern epoch to eliminate the dichotomy and naturalize technology, the nature-technology divide broadly persists in the above definition. In a certain sense, the artificiality definition of synthetic biology presupposes the ongoing plausibility of the Aristotelian concept of nature, neglects the Baconian one, and argues for an epochal break in understanding bio-objects and bio-nature: these are not given, they are made.

as a continuation of further trends already perceived as terrible and dangerous. According to the proponents of this definition, we are experiencing a slight shift and mainly a continuation, not an epochal break; nothing is really new under the sun. Synthetic biology merely extends and complements biotechnology. Drew Endy, a key advocate of synthetic biology, perceives only an "expansion of biotechnology" (Endy 2005, p.449). Similarly, but from a more critical angle, the Action Group on Erosion, Technology and Concentration (ETC) defines "synthetic biology" as an "extreme gene technology", mainly because synthetic biology is based on gene synthesis and cell techniques such as nucleotide synthesis, polymerase chain reaction (PCR), or recombined cloning (ETC 2007). The underlying methods, techniques, and procedures have been well established since the late 1970s. Although there have been tremendous advances from a quantitative standpoint, it is hard to discern any qualitative progress in the core methods.[5] Briefly, this position perceives a continuation in methods—in contrast to a divergence from biology or nature according to the above mentioned two definitions of synthetic biology.

3   Deficits of the three definitions

The three definitions — the engineering, the artificiality, and the extreme biotechnology definition — tell three different stories. Each one exhibits some degree of plausibility and conclusiveness. In spite of their apparent differences, all are concerned (*first*) with disciplinary biology or biological nature and (*second*) with a rational design ideal in conjunction with a specific understanding of technology, technical systems, and engineering action. However, this is not the whole story.

First, the focus on biology as a discipline alone, including a discipline-oriented framing, prevents an exhaustive characterization of the new technoscientific wave. Synthetic biology is much more *interdisciplinary* than disciplinary at its nucleus. This needs to be taken into account when looking for an adequate definition: biologists, computer scientists, physicists, chemists, physicians, material scientists, and people from different engineering sciences are engaged in synthetic biology. Since various disciplinary approaches, methods, and concepts coexist in synthetic biology, the term seems to be a label for a new and specific type of interdisciplinarity (cf. Schmidt

---

5   This definition rarely deals with goals or objects, but with methods and techniques. Its proponents claim (1) that methods constitute the core of synthetic biology, (2) that there has been no breakthrough in the synthetic/biotechnological methods, and, moreover, (3) that a quantitative advancement cannot induce a qualitative one.

2008b). Accordingly, a strong biology bias would surely be overly simplistic and entirely inadequate; to frame synthetic biology merely as a new subdiscipline of biology would represent a far too narrow approach. Thus, we need to ask whether we are faced with a much more fundamental technoscientific wave than simply a change in one particular discipline or academic branch alone.

Second, in line with what has become known as bionano or nanobio research, the three definitions look at synthetic biology from the angle of technology and engineering. This manner of approach appears viable in some respects: synthetic biology extends and complements advancements in nanotechnology and hence spurs a position that can be called "technological reductionism" (Schmidt 2004, pp.35f.; cf. Grunwald 2008, pp.41f./190f.). Technological reductionists aim at eliminating the patchwork of engineering sciences by developing a fundamental technology, or a "root, core, or enabling technology" (Schmidt 2004, p.42). The slogan promoted by technological reductionism is: Shaping, constructing, and creating the world 'atom-by-atom'. Eric Drexler is a prominent advocate of technological reductionism. He argues that there are "two styles of technology. The ancient style of technology that led from flint chips to silicon chips handles atoms and molecules in bulk; call it bulk technology. The new technology will handle individual atoms and molecules with control and precision; call it molecular technology" (Drexler 1990, p.4). Now, it has been argued that the three definitions of synthetic biology given above concur strongly with technological reductionism. It certainly seems plausible to put synthetic biology in the context of this new type of technology oriented reductionism. But whether that is all that can, or should, be said to characterize synthetic biology still needs to be clarified. Most clearly, synthetic biology differs from nanotechnology, which can be regarded as a paradigm of a technological reductionist approach. Synthetic biology claims to pursue an approach that is complementary to nanotechnology and has been called 'systems approach' or, in a more visionary sense, 'holistic'. Given the widespread reference to 'system', including the claim of successful application of 'systems thinking', synthetic biology seems to involve a convergence, or dialectical relationship, of seemingly contradictory concepts: (systems') holism and (technological) reductionism (with its strong control ambitions and emphasis on rational engineering). This inherent dialectic is obviously central to an adequate and appropriate understanding of synthetic biology. The three definitions presented so far do not consider this point.

4 Towards a new technoscientific paradigm?

For a better and more comprehensive characterization of synthetic biology we should not restrict ourselves to goals (as in definition 1), objects ('ontology', as in definition 2), or methods ('methodology', as in definition 3), but also consider the underlying principles, concepts and theories within the technoscientific field. A further definition is prevalent in synthetic biology's R&D programs: *the self-organization* (or systems) *definition*. Synthetic biology harnesses, or at least aims to harness, self-organization power (of nature) for technological purposes.

The paradigm of self-organization is present in many papers on synthetic biology: "Harnessing nature's toolbox" in order to "design biological systems", as David A. Drubin, Jeffrey Way, and Pamela Silver (2007) state. As early as in 2002, before synthetic biology had been coined as a term (although its main ideas were already present), Mihail Roco and William Bainbridge anticipated new frontiers in research and development by "learning from nature". They perceived the possibility of advancing technology by "exploiting the principles of automatic self-organization that are seen in nature" (Roco and Bainbridge 2002, p.258). According to Alain Pottage and Brad Sherman, the basic idea of synthetic biology is to "turn organisms into manufactures" and to make them "self-productive" (Pottage and Sherman 2007, p.545). "We think that in order to design products 'of biological complexity' that could make use of the fantastic fabrication abilities […], we must first liberate design by discovering and exploiting the principles of automatic self-organization that are seen in nature" (Pollack 2002, p.161). In a similar vein, the 2009 report "Making Perfect Life" of the European TA Group refers to advancements in synthetic biology: "Synthetic biology […] present[s] visions of the future […]. Technologies are becoming more 'biological' in the sense that they are acquiring properties we used to associate with living organisms. Sophisticated 'smart' technological systems in the future are expected to have characteristics such as being self-organizing, self-optimizing, self-assembling, self-healing, and cognitive" (ETAG 2009, p.4).[6] Obviously, "[t]he paradigm of complex, self-organizing systems is stepping ahead at an accelerated pace, both in science and in technology" (Dupuy 2004, pp.12f.; cf. Luisi and Stano 2011).

The systems approach of putting the self-organization power of bioengineered entities at the very center of the new technoscientific wave has enjoyed an impres-

---

6   And the ETAG goes on to stress: "Central in their ideas is the concept of self-regulation, self-organization and feedback as essential characteristics of cognitive systems since continuous adaption to the environment is the only way for living systems to survive" (ETAG 2009, p. 25).

sive history over the last three decades. It goes back to one of the most popular and highly controversial publications by K. Eric Drexler in the early 1980s. Drexler talks about "self-assembly", "engines of creation", and "molecular assemblers" (Drexler 1990; cf. Nolfi and Floreano 2000). "Assemblers will be able to make anything from common materials without labor, replacing smoking factories with systems as clean as forests." Drexler goes even further and claims that emergent technologies "can help mind emerge in machine." Richard Jones takes up Drexler's ideas and perceives a trend towards "self-organizing [...] soft machines" that will change our understanding of both nature and technology (Jones 2004).

Synthetic biology – this is interesting to note – does not stand alone. Self-organization also plays a constitutive role in other kinds of emerging technologies such as

1. Robotics, AI, ubiquitous computing, autonomous software agents, bots;
2. Nano- and nanobio-technologies;
3. Cognitive and neuro-technologies.

In addition, self-organization in technical systems serves as a leitmotif in science policy: "Unifying science and engineering" seems to become possible by "using the concept of self-organized systems" (Roco and Bainbridge 2002). Self-organization appears to be the kernel of the ideal of the *convergence of technologies*, and it also seems central to any kind of *enabling technology* (ibid.; Schmidt 2004). In oother words, synthetic biology is not unique; it can be considered as just a prominent example of a very universal trend in technology.

5   Late-modern technology

If we take the visionary promises as serious claims, they announce the emergence of a new type of technology. We do not know whether the promises can be fully kept. However, should this be the case, we would encounter a different kind of technology: a *late-modern technology*.

Late-modern technology does not resemble our established perception and understanding of technical systems. It displays nature-like characteristics; it does not appear as technology; it seems to be "un-technical" or "non-artifical"; the signs and signals, the tracks and traces are no longer visible (Hubig 2006; Karafyllis 2003; Kaminski and Gelhard 2014). Technical connotations have been peeled off; well-established demarcation lines are blurred. Late-modern technology seems to possess an intrinsic momentum of rest and movement within itself — not an extrinsic one.

Such characteristics come close to the Aristotelian and common life-world understanding of nature: technology *is* alive or *appears* to be alive, as nature always has been. The internal dynamics (i.e., acting, growing, and changing) of self-organization technology make it hard to draw a demarcation line between the artifactual and the natural in a phenomenological sense. Traditional technical connotations have been peeled off. Nature and technology seem indistinguishable. Even where it is still possible to differentiate between the artificial and the natural, e.g., in robotics, we are confronted with more and more artifacts displaying certain forms of behavior that traditionally have been associated with living systems. The words used by Schelling and Aristotle to characterize nature also seem to apply to technology. Late-modern technical systems are "not to be regarded as primitive." Late-modern technology appears to act by itself: (a) it creates/produces; (b) it selects means to ends (means-ends rationality); (c) it takes decisions and acts according to its environmental requirements. Technology evidently presents itself as an actor: "autonomy" — a term that is central to our thought tradition — seems to be ascribable to these systems.

What is the background of this trend towards a *phenomenological convergence* of nature and technology or, in other terms, towards a *phenomenological naturalization of technology* — besides and in addition to "technological reductionism" (Schmidt 2004)? Much more relevant and foundational, it seems, is what could be called *nomological convergence* that gives rise to a fundamental trend towards a *nomological naturalization of technology*. Mathematical structures that describe self-organization in technical systems are said to converge with those in nature. Although the objects might differ, their behavior and dynamics show a similarity. According to M.E. Csete and J.C. Doyle, "advanced technologies and biology are far more alike in systems-level organization than is widely appreciated" (Csete and Doyle 2002, p.1664). The guiding idea of nomological convergence can be traced back to the cyberneticist Norbert Wiener (1968, first published in 1948). He defined structure-based convergence with regard to specific "structures that can be applied to and found in machines and, analogously, living systems." As the physicist, philosopher and programmatic thinker Carl Friedrich von Weizsäcker pointed out about 50 years ago: "Structural sciences encompass systems analysis, information theory, cybernetics, and game theory. These concepts consider structural properties and features of different objects regardless of their material realm or disciplinary origin. Time-dependent processes form a common umbrella that can be described by an adequate mathematical approach and by using the powerful tools of computer technology"[7] (Weizsäcker 1974, pp.22f.; cf. Küppers 2000; Schmidt 2008a).

Today, we can add self-organization theories which encompass nonlinear dynamics, complexity theory, chaos theory, catastrophe theory, synergetics, fractal geometry, dissipative structures, autopoiesis theory, and others. Following the first wave of structural and systems sciences such as information theory, game theory, and cybernetics (e.g., Bertalanffy, Wiener, Shannon, von Neumann) in the 1930s and 1940s, we are now experiencing a second wave (e.g., Maturana, Varela, Prigogine, Haken, Foerster, Ruelle, Thom) that began in the late 1960s. Self-organization, macroscopic pattern formation, emergent behavior, self-structuring, growth processes, the relevance of boundary conditions, and the Second Law of Thermodynamics (entropy law) with its irreversible arrow of time are regarded as conceptual approaches to disciplinarily different types of objects, based on evolutionary thinking in complex systems. Assisted by the spread of computer technology, concepts of self-organization had a tremendous impact on scientific development in the second half of the 20th century.

6 Self-organization and instability

Since Kant and Schelling, the concept of self-organization has been in flux. However, 'self-organization' seems to have retained its central meaning, which is the immanent creation and construction of *novelty*: the emergence of novel systemic properties — new entities, patterns, structures, functionalities, capacities. Beyond the philosophical dispute on the notion and characteristics of novelty, the following criteria to specify 'self-organization' are widely accepted (Stephan 2007; Schmidt 2008a):

1. Internal dynamics, inherent processes, and time-dependency;
2. Irreducibility of the description length;
3. Unpredictability of the self-organizing or emergent phenomena.

In consequence, self-organization processes cannot be generally separated from their environment; they are hard to control by an external actor. "The engineers of the future will be the ones who know that they are successful when they are surprised by their own creations" (Dupuy 2004). In brief, the notion of self-organization is,

---

7   My translation from German (J.C.S.).

from an engineering perspective, linked to characteristics such as 'productivity', 'processuality', and 'autonomy'.

It has been said that synthetic biology's core is its claim of harnessing self-organizing power for technological purposes. But what is the core or root of self-organization? The basic answer that I propose is that *instabilities* turn out to be essential for self-organization; they are constitutive to all systems or structural theories (Schmidt 2008a). The physicist J.S. Langer (1980), for instance, underlines the role of "instabilities for any kind of pattern formation." According to Werner Ebeling and Reiner Feistel (1994, p.46), "self-organization is always induced by instability of the 'old' structure through small fluctuations. This is why studying instability is of major importance." Gregory Nicolis and Ilya Prigogine (1977, pp.3f.) argue that "instabilities [are ...] necessary conditions for self-organization." Wolfgang Krohn and Günter Küppers (1992, p.3), in the same vein, emphasize that "instabilities are the driving force and the internal momentum for systems evolution and development."

Instabilities can generally be regarded as situations in which a system is on a razor's edge: criticalities, flip or turning points, thresholds, watersheds. They generate sensitive dependencies, bifurcations, phase transitions. The classic-modern *strong* type of causation does not govern these processes; rather, it is the *weak* type of causation that enables feedback procedures and amplification processes. Instabilities can induce random-like behavior, deterministic chance, and law-based noise, which are inherently linked to uncertainty. The most prominent example used to illustrate instability is the butterfly effect. The beating of a butterfly's wings in South America can have tremendous influence on the weather in the U.S. and cause a thunderstorm.[8]

Unstable systems show certain limitations of: predictability, reproducibility, testability, and reductive describability. An isolation or separation from their environment is impossible as they interact with it continuously. In general, instability should not be equated with the collapse of a system. Insofar as engineers today aim at harnessing self-organization power, they have to provoke and stimulate insta-

---

[8] The list of examples is extensive (cf. Schmidt 2011): the emergence and onset of a chemical oscillation, the role dynamics of a fluid in heat transfer, an enzyme kinetic reaction, a gear chattering, or turbulence of a flow. A fluid becomes viscous, ice crystallization emerges, a phase transition from the fluid to a gas phase takes place, a solid state becomes super-fluid, a laser issues forth light, a water tap begins to drip, a bridge crashes down, an earthquake or tsunami arises, a thermal conduction process comes to rest, and a convection sets in, e.g., Bénard instability. New patterns and structures appear. These examples underscore the fact that instabilities are the necessary condition for *novelty*. The various definitions of complexity refer directly or indirectly to instabilities.

bilities: self-organization requires that a system's dynamics pass through unstable situations. To put it metaphorically: late-modern technology can be considered the technoscientific attempt to stimulate/initiate a *dance on the razor's edge*.

7 Inherent ambivalence

Instability-based technology is ambivalent because it carries an internal conflict or considerable dialectic. On the one hand, instabilities constitute the core of self-organization and, hence, of technologically relevant self-productivity. On the other hand, instabilities are linked with limitations with regard to the construction and design of the technical systems and also with regard to the possibility of subsequently monitoring and controlling them (cf. Köchy 2011; Schmidt 2012b).

A closer examination can help us to appreciate why it is so difficult to engineer and harness self-organization for technological purposes — and why "rational design" and "rational engineering" are limited (Giese et al. 2013). When instabilities are present, tiny details are of major relevance; minor changes in some circumstances can cause tremendous, unforeseeable effects. Such systems cannot be separated from their environment. Unstable systems lack predictability, reproducibility, and separability. A clear input-output mechanism — as in the classic mechanistic and means-oriented understanding of a technical system — does not exist here. Tiny details and perturbations are hard to control, due to empirical-practical and to fundamental-principle uncertainties. Therefore, reasonable concerns can be raised as to whether, e.g., "synthetic biology will enable the design of 'biological systems' in a rational way" (European Commission 2005). Thus, instabilities challenge the classic rational attempts (a) to intervene and manipulate (given) self-organizing systems, (b) to construct and design such (new) systems, and further, (c) to control and monitor them.

Due to these limitations, technology and instability were, traditionally, like fire and ice. According to the classic-modern view of technology, instabilities ought to be excluded from technology. If instabilities occurred, the traditional objective was to eliminate them. Constructability and controllability, including a clear input-output relation, are only guaranteed when stability exists. Technology was traditionally equated with and defined by stability.[9] These thoughts concur

---

9   The fundamental properties of such a late-modern technology have the power to change the world we live in. "Because engineered micro-organisms are self-replicating and capable of evolution," as Tucker and Zilinskas argue, "they belong in a different risk

with what Alfred Nordmann perceives as a "limit [that] could be reached where engineering seeks to exploit surprising properties that arise from natural processes of self-organization" (Nordmann 2008, p.175). "No longer a means of controlling nature in order to protect, shield, or empower humans, technology dissolves into nature and becomes uncanny, incomprehensible, beyond perceptual and conceptual control" (ibid., p.173). We are on the way to "surrender[ing] control to pervasive technical systems" (ibid., p.182). As Jean-Pierre Dupuy puts it, "the novel kind of uncertainty that is brought about by those new technologies is intimately linked with their being able to set off complex phenomena in the Neumannian sense" (Dupuy 2004, p.10). Non-knowledge, ignorance, and uncertainty are co-produced with the productiveness of the late-modern technical systems; they are inherent by-products and do not simply emerge in the societal context of diffusion, use, and consumption. Instability-based technology has a life of its own.

8 Precursors: Jonas and Luhmann

As early as in the 1980s, Hans Jonas (1985) anticipated the central characteristics and the limits of "engineering biology" — which today can be regarded as an example of late-modern technology (Köchy 2012; Schmidt 2012b). He diagnosed a historically new technoscientific era and drew a dividing line between the classic engineering type of technology ("engineering art") and the new type. As Jonas argued, this new type of technology differs in a qualitative way from our established understanding of what technology is or could be. "In th[is latter ...] case of dead substances, the constructor is the one and only actor with respect to a passive material [= classic-modern technology]. [In contrast, in the case of the] biological organism, activity meets activity: biological technology is collaborative with the

category than toxic chemicals or radioactive materials" (Tucker and Zilinskas 2006). Indeed, this objection already applies to classic substances of biotechnology. But the related challenges in the realm of synthetic biology go much deeper and are to be regarded as more pressing. In particular, the *principle of similarity (and resemblance)* that constitutes the backbone of any risk assessment cannot be applied to most substances and tissues of synthetic biology. This principle is based on the assumption that if a new (bio-) system has some similarity to one that is known, the new system will behave similarly to the well-known one and exhibit essentially similar properties. Most self-organizing bio-systems are not similar, owing to their intrinsic instability, and therefore they cannot be compared to other bio-systems: the principle of similarity is not applicable, due to the onto-technological core of late-modern technology.

self-activity of an active 'material'" (Jonas 1985, p.165). Jonas lists the following characteristics of this new type of technology:

1. collaborativeness, self-activity, autonomy;
2. irreversibility and historicity;
3. complexity, time-dependency, evolution, and limitation of predictability;
4. individuality, non-experimentability and obstacles regarding reproducibility;
5. interactive causation as a different kind of causality.

However, Jonas did not take his clear phenomenological description any further. Based on the argumentation developed in this paper, it is not only the organismic that constitutes the central difference, but also, and above all, instability-based self-organization. Jonas' *philosophy of technology* is much more fundamental than Jonas himself seems to have assumed. The explication and exploration of the new type of technology could also substantiate Jonas' formulation of his *imperative of responsibility*, including precautionary principles (Jonas 1984; cf. Schmidt 2013).

Niklas Luhmann's rich, but not yet fully acknowledged *philosophy of technology* was also concerned with the ambivalence rooted in technical (co-)produced instabilities. Although his approach differs considerably from Jonas', both have much in common. Luhmann was skeptical whether a late-modern type of technology is — as 'technology' or as a 'technical system' — possible at all. Based on his understanding of technology as "functional simplification" or "simplifying isolation," he anticipated problems that are inherent *to*, and characteristics *of*, late-modern technology — although he developed his thinking by referring *not* to late-modern technology but to modern high-technology, e.g., nuclear energy.[10] In high-technology we are confronted with "the chaos problem, the interference problem and the problem of singular, random-like events" (Luhmann 2003, p.100). Luhmann is here drawing, obviously, on features that are induced by instabilities. According to him, the problems originate from the "form of the technical systems themselves that constitutes the border between the included and excluded causalities." The "form" of high-technical systems — as well as of late-modern technical systems — renders it impossible to keep these two kinds of causalities apart or, at least, to control them; the non-existence of such a clear border and the non-separability of the two causalities are a consequence of instabilities. Therefore, the chaos problem, the interference problem, and the problem of singular events "destroy the option of a technological regulation of technical systems" (ibid., pp.99f.). Interestingly, the

---

10   This is why 'late-modern technology' should be considered not only as a term referring to technoobjects, but also as a reflexive term.

difference between classic-modern and late-modern technologies is that in the latter instabilities are produced intentionally whereas in the classic-modern account of technology instabilities are highly undesired. Luhmann himself sticks to the classic-modern ideal of technology and questions whether late-modern technology can be conclusively called 'technology' or, more fundamentally, whether a late-modern technology is "as a technical system technically possible at all" (ibid., p.100).

9 Summary: Characterizing late-modern technology

Synthetic biology can be regarded as a prominent example of late-modern technology; however, it is not unique. Another most illustrative example is that of *autonomous stock trading agents*, which are implemented as artificial neuronal networks and genetic algorithms and trained with real 'big data' in the context of application. They are widely spread throughout the global neo-capitalist finance market in connection with *high frequency trading* (HFT) at the stock exchange. Instability-based cascade effects can be induced by the autonomous trader agents — the seemingly uncontrollable algorithmic 'market makers'.

In general, late-modern technology differs from the classic-modern type of technology with regard to three main characteristics. First, *phenomenological characteristics*: late-modern technical systems are based on self-organization. They appear to be un-technical and non-artificial — and also show 'autonomous' behavior and agency properties. The signs and signals, tracks and traces of technology are no longer visible. Culturally established borders are becoming blurred. This universal trend is leading towards a *phenomenological naturalization of technology*. Second, *nomological or ontological characteristics*: the nomological core of late-modern technology is instability — as a necessary condition of self-organization. Instabilities are intentionally built into the form of technical systems and their material structure. Here, we can perceive a trend that could be called *nomological naturalization of nature*. Third, *methodological, epistemological and action-theoretical characteristics*: late-modern technology can be described by the absence of further criteria, it hardly (a) separable from its environment and from the context of application; it lacks (b) reproducibility, (c) predictability, and (d) testability/describability; it gives rise to limitations of (e) constructing and creating; it eludes (f) monitoring and controlling.

Therefore, this kind of technology has or would have, if realized to its full extent, "a life of its own" (Nordmann 2008). It could be regarded as a 'naturalized technology', in other words: as a *phenomenological* as well as a *nomological naturalization of technology*. Whether late-modern technology can be conclusively called 'technology' and is "as a technical system technically possible at all" remains an open question (Luhmann 2003, p.100). Nevertheless, technical systems, devices, things, and objects that are based on instabilities and show self-organizing phenomena are beginning to populate our life-world. From a societal perspective we need to address this instability-based late-modern type of technology and undertake the task of developing procedures to either restrict and contain, or to shape and deal with it.

References

Csete, M.E., & Doyle, J.C. (2002). Reverse Engineering of Biological Complexity. *Science, 295*, 1664-69.

Deplazes, A., & Huppenbauer, M. (2009). Synthetic organisms and living machines: Positioning the products of synthetic biology at the borderline between living and nonliving matter. *Systems and Synthetic Biology, 3*, 55-63.

DFG, acatech, & Leopoldina (2009). Deutsche Forschungsgemeinschaft (German Research Foundation), acatech – Deutsche Akademie der Technikwissenschaften (German academy of technological sciences), Leopoldina – Deutsche Akademie der Naturforscher (German academy of natural scientists). *Synthetic Biology: Positions.* Weinheim: Wiley VCH.

Drexler, K.E. (1990). *Engines of Creation: The Coming Era of Nanotechnology.* Oxford: Oxford University Press.

Drubin, D.A., Way, J.C., & Silver, P.A. (2007). Designing biological systems. *Genes & Development, 21*, 242-254.

Dupuy, J.P. (2004). Complexity and Uncertainty. A Prudential Approach to Nanotechnology. In: European Commission, *Nanotechnologies: A Preliminary Risk Analysis on the Basis of a Workshop Organized in Brussels on 1-2 March 2004 by the Health and Consumer Protection Directorate General of the European Commission* (pp.71-94). http://ec.europa.eu/health/ph_risk/documents/ev_20040301_en.pdf. Accessed: 1 May 2014.

Ebeling, W., & Feistel, R. (1994). *Chaos und Kosmos: Prinzipien der Evolution.* Heidelberg, Berlin: Spektrum.

Endy, D. (2005). Foundations for engineering biology. *Nature, 438*, 449-453.

ETAG (2009). *Making a perfect life. Bioengineering in the 21st century.* European Technology Assessment Group, Rathenau Institute. The Hague.

ETC (2007). ETC Group. *Extreme Genetic Engineering. An Introduction to Synthetic Biology.* http://www.etcgroup.org/sites/www.etcgroup.org/files/publication/602/01/synbiore-portweb.pdf. Accessed: 1 May 2014.

European Commission (2005). *Synthetic Biology – Applying Engineering to Biology. Report of a NEST-High Level Expert Group EUR 21796.* Luxembourg: Office for Official Publications of the European Communities.

Gibson, D.G., Glass, J.I., Lartigue, C., Noskov, V.N., Chuang, R.Y., Algire, M.A., ... & Venter, J.C. (2010). Creation of a bacterial cell controlled by a chemically synthesized genome. *Science, 329*(5987), 52-56.

Giese, B., Koenigstein, S., Wigger, H., Schmidt, J.C., & Gleich, A. v. (2013). Rational engineering principles in synthetic biology: A framework for quantitative analysis and an initial assessment. *Biological Theory, 8*(4), 324-333.
Grunwald, A. (2008). *Auf dem Weg in eine nanotechnologische Zukunft. Philosophisch-ethische Fragen*. Freiburg: Alber.
Grunwald, A. (2012). Synthetische Biologie als Naturwissenschaft mit technischer Ausrichtung. Plädoyer für eine Hermeneutische Technikfolgenabschätzung. *Technikfolgenabschätzung—Theorie und Praxis, 21*(2), 10-15.
Hubig, C. (2006). *Die Kunst des Möglichen: Technikphilosophie als Reflexion der Medialität* (Vol. 1). Bielefeld: transcript.
Jonas, H. (1984). *The Imperative of Responsibility. In Search of an Ethics for the Technological Age*. Chicago: University of Chicago Press.
Jonas, H. (1985). Laßt uns einen Menschen klonieren: Von der Eugenik zur Gentechnologie. In: H. Jonas, *Technik, Medizin und Ethik: Praxis des Prinzips Verantwortung* (pp.162-203). Frankfurt/Main: Insel.
Jones, R. (2004). *Soft Machines*. Oxford: Oxford University Press.
Kaminski, A., Gelhard, A. (eds.) (2014). *Zur Philosophie informeller Technisierung*. Darmstadt: WBG.
Karafyllis, N. (ed.) (2003). *Biofakte*. Paderborn: Mentis.
Köchy, K. (2011). Konstruktion von Leben? Herstellungsideale und Machbarkeitsgrenzen in der Synthetischen Biologie. In: V. Gerhardt, K. Lucas, G. Stock (eds.), *Evolution. Theorie, Formen und Konsequenzen eines Paradigmas in Natur, Technik und Kultur* (pp.233-242). Berlin, Heidelberg: Akademie Verlag.
Köchy, K. (2012). Sind die Überlegungen von Hans Jonas zum Sonderstatus biologischer Technik angesichts der Entwicklung in der Synthetischen Biologie noch haltbar? In: M.B. Bondio, H. Siebenpfeiffer (eds.), *Konzepte des Humanen. Ethische und kulturelle Herausforderungen* (pp.81-101). Freiburg, München: Alber.
Krohn, W., & Küppers, G. (eds.) (1992). *Selbstorganisation. Aspekte einer wissenschaftlichen Revolution*. Braunschweig: Vieweg.
Küppers, B.-O. (2000). Die Strukturwissenschaften als Bindeglied zwischen Natur- und Geisteswissenschaften. In: B.-O. Küppers (ed.), *Die Einheit der Wirklichkeit. Zum Wissenschaftsverständnis der Gegenwart* (pp.89-110). München: Fink.
Langer, J.S. (1980). Instabilities and Pattern Formation. *Reviews of Modern Physics. 52*, 1-28.
Luhmann, N. (2003). *Soziologie des Risikos*. Berlin, New York: de Gruyter.
Luisi, P. L., & Stano, P. (2011). Synthetic biology: Minimal cell mimicry. *Nature Chemistry, 3*(10), 755-756.
Nicolis, G., & Prigogine, I. (1977). *Self-Organization in Nonequilibrium Systems. From Dissipative Structures to Order through Fluctuations*. New York: Wiley.
Nolfi, S., & Floreano, D. (2000). *Evolutionary Robotics: The Biology, Intelligence, and Technology of Self-Organizing Machines*. Cambridge: MIT Press.
Nordmann, A. (2008). Technology Naturalized. A Challenge to Design for the Human Scale. In: P.E. Vermaas, P. Kroes, A. Light, S. Moore (eds.), *Philosophy and Design: From Engineering to Architecture* (pp.173-184). Heidelberg, New York: Springer.
Pollack, J. (2002). Breaking the Limits on Design Complexity. In: M.C. Roco, W.S. Bainbridge (eds.), *Converging Technologies for Improving Human Performance*. Arlington: National Science Foundation.

Pottage, A., & Sherman, B. (2007). Organisms and manufactures: on the history of plant inventions. *Melbourne University Law Review, 31,* 539–568.
Pühler, A., Müller-Röber, B., & Weitze, M.-D. (eds.) (2011). *Synthetische Biologie. Die Geburt einer neuen Technikwissenschaft.* Berlin, Heidelberg: Springer.
Roco, M.C., & Bainbridge, W.S. (eds.) (2002). *Converging Technologies for Improving Human Performance.* Arlington: National Science Foundation.
Schmidt, J.C. (2004). Unbounded Technologies: Working through the Technological Reductionism of Nanotechnology. In: D. Baird, A. Nordmann, J. Schummer (eds.), *Discovering the Nanoscale* (pp.35-50). Amsterdam: IOS Press.
Schmidt, J.C. (2008a). *Instabilität in Natur und Wissenschaft.* Berlin: de Gruyter.
Schmidt, J.C. (2008b). Towards a philosophy of interdisciplinarity. An attempt to provide a classification and clarification. *Poiesis & Praxis, 5*(1), 53-69.
Schmidt, J.C. (2011). Challenged by Instability and Complexity. On the methodological discussion of mathematical models in nonlinear sciences and complexity theory. In: C. Hooker (ed.), *Philosophy of Complex Systems* (pp.223-254) (Series Philosophy of Sciences). Amsterdam: Elsevier.
Schmidt, J.C. (2012a). Quellen des Nichtwissens. Ein Beitrag zur Wissenschafts- und Technikphilosophie des Nichtwissens. In: N. Janich, A. Nordmann, L. Schebek (eds.), *Nichtwissenskommunikation in den Wissenschaften: interdisziplinäre Zugänge* (pp.93-124). Frankfurt a.M.: Lang.
Schmidt, J.C. (2012b). Selbstorganisation als Kern der Synthetischen Biologie. Ein Beitrag zur Prospektiven Technikfolgenabschätzung. *Technikfolgenabschätzung – Theorie und Praxis, 21*(2), 29-35.
Schmidt, J.C. (2013). Defending Hans Jonas' Environmental Ethics: On the Relation between Philosophy of Nature and Ethics. *Environmental Ethics, 35,* 461-479.
Schwille, P. (2011). Bottom-Up Synthetic Biology: Engineering in a Tinkerer's World. *Science, 333,* 1252-54.
Stephan, A. (2007). *Emergenz: Von der Unvorhersagbarkeit zur Selbstorganisation.* Paderborn: Mentis.
TESSY (2008). Towards a European Strategy for Synthetic Biology. *Synthetic Biology in Europe.* Information leaflet. http://www.tessy-europe.eu/public_docs/SyntheticBiology_TESSY-Information-Leaflet.pdf. Accessed: 23 May 2013.
Tucker, J.B., & Zilinskas, R.A. (2006). The promise and perils of synthetic biology. *New Atlantis, 12*(1), 25-45.
Venter, J.C. (2010). The Creation of 'Synthia' – Synthetic Life. The Naked Scientist, Science Interview May 23, 2010. http://www.thenakedscientists.com/HTML/content/interviews/interview/1332/. Accessed: 20 June 2013.
Weizsäcker, C.F. v. (1974). *Die Einheit der Natur.* München: dtv.
Westerhoff, H.V., & Palsson, B.O. (2004). The evolution of molecular biology into systems biology. *Nature Biotechnology, 22*(10), 1249–52.
Wiener, N. (1968). *Kybernetik: Regelung und Nachrichtenübertragung in Lebewesen und Maschine.* Hamburg: Rowohlt.

Synthetic Biology
On epistemological black boxes, human self-assurance, and the hybridity of practices and values

Oliver Müller

1 Introduction

In 2010, when J. Craig Venter and his group published the experiment in which the DNA of a bacterium had been completely replaced by a synthesized DNA (Gibson et al. 2010), the international press responded with excitement. The headline of the German news magazine DER SPIEGEL illustrated the enthusiasm: "Breakthrough: Scientists create first artificial life"[1] (*Der Spiegel* 2010). Since synthetic biology has been present in the media, the 'usual suspects' or patterns of interpretation come into the picture. For example, the literary figure of Faust is brought up (Hofmann 2010), as is Prometheus (*The Economist* 2006) or Frankenstein (Epping 2008; Belt 2009). Additionally, there is talk of the "homunculus" (H. Müller 2010) as a means to characterize the products of synthetic biology, and of the "golem"[2] – all of them good old reminiscences of the hubris of the homo faber.

A certain unease concerning new, uncontrolled possibilities in the biotechnologies is discernible. In order to differentiate, one has to ask: is synthetic biology taking a step forward compared to 'classical' biotechnology? Considering how synthetic biologists portray themselves/portray their field, synthetic biology seems to be clearly distinguishable from 'traditional' biotechnology. The aim of synthetic biology is the production of cell structures that do not exist in nature. The phrasing of the EU Expert Group on "New and Emerging Science and Technology" (NEST) might count as paradigmatic: "Synthetic biology is the engineering of biology: the synthesis of complex, biologically based (or inspired) systems which display functions that do not exist in nature." (European Commission 2006) Not surprisingly,

---

1   Own translation.
2   Cf. Paul Root Wolpe's statement in (Rejeski et al. 2010).

Venter himself claims that the design of entirely new forms of life is limited only by our imagination (Alleyne 2010).

'Synthetic biology' is usually seen as an umbrella term for several biotechnological methods, approaches, programs, and not least, various bio-political agendas (cf. Bensaude Vincent 2013a; 2013b). But the tendency to use the semantic field of 'creation' points to the paradigm shift effected by synthetic biology: from manipulation to creation. Although the entities produced in synthetic biology and 'classical' genetic engineering might resemble each other, the explicit goal of synthetic biology is to produce *new* forms of life. In basic research, one speaks of 'orthogonality' when alternative bio-chemical systems are 'constructed' by using 'atypical elements'. Witty biotechnologists refer to this as 'parallel worlds' or 'parallel universes'. From a philosophical point of view, one can identify characteristic sets of technoscientific strategies, epistemological problems, and ontological challenges regarding hybrids such as 'bio bricks' and 'artificial cells', 'bio-artifacts', etc.

In order to make an adequate assessment of synthetic biology, which can be the basis for an ethical assessment of this research field, one must look not only at exactly how synthetic biologists work, what techniques they use and what their objectives are, but also at the understanding of 'life' and 'nature' on which their research is based – and how the understanding of nature and human self-understanding might change when the production of 'new' organisms is the stated goal of this research field. One famous example is the iGEM contest at MIT (www.igem.org). Every year, students are invited to synthesize new organisms. Often, it is not the aspect of 'usefulness' that is rewarded, but the 'originality' to build up a new and fancy microorganism – microorganisms that some day may serve as a basis for profitable organic products. Synthetic biology presents itself not only as a science aiming to extend knowledge within its field or to explore what is 'useful' and 'healing' for severe diseases. Synthetic biology is also a discipline into which decidedly 'ontological goals' are written (when 'ontology' is very roughly understood as the scope of our understanding of forms of being and in particular, in this context, of the relation and status of artifacts and living beings) (cf. Schyfter 2011).

In the following, I focus on some of the epistemological, ontological, and anthropological aspects that are relevant for the discussion. Being fully aware of the problem with defining synthetic biology, I assume that synthetic biology is mainly based on an engineering paradigm or 'agenda' – a radicalized idea of manufacturing and of technically constructed living entities. This has its effects on the terminology: biological substrates, in particular cells, are described using partly crude machine metaphors such as 'living machines' or 'bio bricks'. It is no longer the 'manipulation' of microorganisms or genes, but 'synthesizing life' or 'creation of life' – a common expression, as the above-cited paper by Venter indicates.

Discussing the goals and needs of synthetic biology is important because the constitution and program of the biotechnologies provide the framework and thus set the course for developments. The self-portrayal of synthetic biologists and the definition of their subject are therefore never just a simple representation of what is or what is done, and not only clever PR – but already a form of politics and bio-politics.

I will contribute to the debate on synthetic biology by showing that epistemological black boxes correspond with a certain tradition of human self-assurance via technological capabilities. I will argue that the terminology of 'creation' has its origin in the intricate combination of epistemological black boxes and anthropological dimensions that underlie scientific processes and shape synthetic biology in particular. One of the consequences of this 'amalgamation' is the genesis of what can be called 'hybrid practices', a conglomeration of different types of practices in the field of synthetic biology and attitudes to living entities. I close my reflections with an outlook on the necessity of a new theory of values in biotechnology.

2 Epistemological black boxes

In the context of synthetic biology, obscure ontological chimeras are afloat. "Artificial cells," "bio bricks," "chassis," and even "living machines" are leading metaphors (Tucker and Zilinskas 2006, p.25; *The New Atlantis* 2004, p.101; Ball 2004; Benner and Sismour 2005; Luisi 2002; Luisi et al. 2006, p.1). Living things are apparently sought to be explained in mechanistic terms – a well-known method from the mechanistic tradition, of which Descartes and LaMettrie were a part. And yet, such terms and metaphors are not only forms of knowledge, they also imply the 'appropriate' way to deal with those biotic entities.

At first glance, one might interpret the invention of such ontological chimeras as an indication of the new possibilities in synthetic biology to produce new life forms, and, hence, of the new 'ontological status' of these biotic products. So, what could be more adequate than to coin new terms? On closer inspection, however, it becomes clear that one is dealing with a kind of ontological 'magic trick' that can be exposed by showing that genesis and product are confused. If the ontological chimeras denote the technical process from which they are derived, then these entities are simply living beings. Likewise, it would be nonsense to refer to a child conceived in vitro as 'artificial life'. Nevertheless, it must be noted at this point that one can only figure out the concept of being that is behind the life sciences if one differentiates the epistemological, the sociological, and the technoscientific entanglements, which are combined in the 'laboratory systems' or 'experimental systems' (see, as locus classicus, Rheinberger (1997)). That is to say, what is called a 'hard' fact depends ultimately on a "style of thinking", to use an expression by Ludwik Fleck (1979).

Regarding the highly artificial technologies, one has to query on a theoretical level – before all production intent – the extent to which one is dealing with living beings and not with rather special artifacts.

This means that – for the synthetic biologist as well as for any scientist – 'living' objects can be neither independent of the cognitive process and the specific 'order of the laboratory' nor just a result of the manufacturing process. Hence, one can argue that the use of the term 'hybrid' is more appropriate to capture the peculiar status of the entity observed in the laboratory and to detect the peculiarity of the produced life forms. One must therefore reject a naïve conceptualization of the microorganic as a mere 'object'. One is dealing with a complex mixture of the technical, natural, and social sphere (Calvert 2010; Knorr-Cetina 2009; see also Latour 1987; 1993). In consequence, the discourse is dominated by the excessive use of metaphors which awkwardly attempt to address the status of the entities in the laboratory procedures.

One may be tempted to reduce the complexity – yet the 'hybrid' in the theory of science is not a living machine. Even if the transformation of metaphors in terminology is common in the life sciences and is giving rise to new object spheres with new object functions and object properties, one must exercise caution. One example for the intricate use of metaphors is the semantic field of 'reading' and 'writing', 'information', and 'communication' which is used to build concepts of cellular structures and genetic processes. At this point, mention should be made of the studies of the historian of science Evelyn Fox Keller (1995; 2002). Metaphors can be very helpful to understand life processes, but these social and scientific constructs can also lead to problematic 'ontologizations'.

Recent comments on the ontological status of the products of synthetic biology, for example by Anna Deplazes (Deplazes and Huppenbauer 2009) or Marianne Schark (2012, p.23), stress that the boundary between organism and machine is still to be sharply drawn. The ontological blurring is normatively important since ethical values are always derived from the distinction between organism and machine. Moral values may be judged as minimal in microorganisms, but a diffuse 'self' is ascribed even to cell cultures that show only a few 'vital' functions in order to distinguish them from artifacts. Karin Knorr Cetina has compiled the 'relations' between lab 'stuff' and microorganisms in interviews with laboratory technicians (Knorr-Cetina 1981).

On the scale of things, biotechnologies form biotic substrates and manipulate or 'reprogram' them – to take up a computer metaphor also prominent in this context

– but do not constitute a new object area between nature and technology. Thus, if synthetic biology cannot create new entities between nature and technology after all, are synthetic biologists then really able to create new life?

It is not surprising that the judgment as to whether new life or new life forms are created by synthetic biology depends on the definition of life. The philosopher Andreas Brenner (2007, p.159) observed some "definitional requirements that bring the possibility of synthesizing 'life' within reach."[3] He has in mind the well-known definition by the biotechnologist and chemist Steve A. Benner, who says: "Any chemical system that combines these properties [– the ability to undergo spontaneous transformation, the ability to direct the synthesis of self-copies –] will be able to undergo Darwinian selection, evolving in structure to replicate more efficiently. In a word, 'life' will have been created" (Benner 2003, p.118).

Regarding the above quote, one can observe a certain 'sovereignty of definition' concerning the concept of life and its use in synthetic biology. This is not trivial, because technologies and changing laboratory situations also imply changes in the concept of life. One has the impression that the reduced definition of life in terms of biochemical criteria only allows a certain and limited hypothesis about living matter – one that is sufficient for the corresponding laboratory requirements. The reduced definition of life in terms is part of a certain experimental system. Experimental settings typically make implicitly or explicitly use of reduced definitions of what life is. And this reductionism has epistemological reasons: One has to be aware that one is dealing with "epistemic things" (in Rheinberger's sense). Consequently, it is necessary to distinguish between criteria used to describe epistemic things in the labs and 'life itself'. Regarding the individually tailored reduced definition of 'life', Benner's "proof of life" is consistent – but nevertheless only part of his specific experimental set-up. As a side note, one can provocatively say that the specialness of living matter proves to be resistant to laboratory determinations. Furthermore, the 'core' essence of living matter (whatever that might be) escapes our technologies.

However, creating life de novo is not possible. Benner's euphoric "'life' will have been created" is based on a confusion of result and genesis of the life form. Synthesizing technologies are artificial, the components are 'natural' from the beginning to the end of the biotechnological processes. This does not justify addressing the procedure as 'creation' of life. Drawing on a Kantian distinction, one can say that the notion of life in synthetic biology is "analytical", because it relies merely on the criteria of standard definitions. This means: we only identify living entities by referring to a feasible definition of 'life'. But defining and subsuming is not creating.

---

3   Own translation. German original: "Gleichwohl fallen einige definitorische Festlegungen auf, die die Möglichkeit, 'Leben' zu synthetisieren, in greifbare Nähe bringen."

Nevertheless, one finds excessive borrowing from the semantic field of the 'creative', the 'new', and the 'novel' in the scientific agenda of the discipline, from its programmatic design to the doctoral theses of young researchers. In search of the origins of such terminology, one will come across Richard Feynman. The physicist's famous quote "What I cannot create, I do not understand"[4] is found in many self-portrayals of synthetic biologists. The conjunction of knowing and producing has a long tradition, from Giambattista Vico's "verumet et factum convertuntur" and Kant's *Critique of Pure Reason* to scientific practices in the 20th century.

But the Feynman dictum could be an epistemological trap: we gain knowledge in designing synthesizing processes and also understand our producing technologies and the experimental settings – but we do not understand 'life'. The confusion stems from the following: the 'creation' of synthesizing technologies and experimental settings in order to produce new genome sequences definitely helps one to 'understand' the new, but not life per se. As mentioned above, one is dealing in laboratories with ontological hybrids and not with 'life' or 'nature' itself.

From attempts to 'create' life forms one can only learn something about the functionality of the technologies and observation methods. It is a fallacy to deduce knowledge of what life is from knowledge about the synthesizing technologies. One can contradict Feynman by stating: engineering life is not the same as understanding life. Alfred Nordmann (2015, p.36) identifies "robust black boxes" in the creation processes and rational strategies of synthetic biology: "These, to be sure, are rational strategies by which to work around limits of knowledge, and to achieve technical solutions in the absence of information about mechanical detail. In other words, these are rational strategies to create robust black boxes or modules." Or, as Holger Breithaupt (2006, p.22) puts it polemically: "In fact, ignoring the unknown is a main idea behind synthetic biology."

3 Anthropological implications: demiurgic self-assurance

At first glance, it seems strange that the 'creative' is essential for the agendas of synthetic biology. As studies show, however, the human being's creativity has been a core dimension in the development of technologies – at least since the Modern Ages when the ancient and medieval paradigms of 'mimesis' were questioned in

---

[4] See, as just one typical example, Sven Panke's report (2008, p.1). See also Markus Schmidt's commentary (2009).

the light of a new anthropology. The human self-perception changed; man and his technology were seen as the source for new 'worlds' and alternatives to the old metaphysical order. Man started to explore the bounds of possibility and contingency as a matrix for the development of technologies. It is not surprising that he also considered the creation of life or life forms such as animals.

With Francis Bacon and Immanuel Kant there are two opposing positions in the discourse. In his famous science utopia *New Atlantis*, Bacon imagined a world in which the creation of living entities is a self-evident part of scientific progress:

> We have also parks and enclosures of all sorts of beasts and birds which we use not only for view or rareness, but likewise for dissections and trials; that thereby we may take light what may be wrought upon the body of man. Wherein we find many strange effects; as continuing life in them, though divers parts, which you account vital, be perished and taken forth; resuscitating of some that seem dead in appearance; and the like. We try also all poisons and other medicines upon them, as well of chirurgery, as physic. By art likewise, we make them greater or taller than their kind is; and contrariwise dwarf them, and stay their growth: we make them more fruitful and bearing than their kind is; and contrariwise barren and not generative. Also we make them differ in colour, shape, activity, many ways. We find means to make commixtures and copulations of different kinds; which have produced many new kinds, and them not barren, as the general opinion is. We make a number of kinds of serpents, worms, flies, fishes, of putrefaction; whereof some are advanced (in effect) to be perfect creatures, like be[a]sts or birds; and have sexes, and do propagate. Neither do we this by chance, but we know beforehand, of what matter and commixture what kind of those creatures will arise. We have also particular pools, where we make trials upon fishes, as we have said before of beasts and birds. We have also places for breed and generation of those kinds of worms and flies which are of special use; such as are with you your silkworms and bees [...] (Bacon 1841, p.267).

The main idea behind the vision of creation is the usefulness of these creatures. Nature is optimized in order to serve human needs. Bacon's vision is an example of the correspondence between technological promises and human self-assurance – and it is not by chance that one finds this correspondence embedded in the story of self-created creatures, specially tailored to the structure and the needs of future civilization. Incidentally, it is remarkable that Bacon envisioned only the creation and manipulation of lower animals. Bacon's text can be read as one with a hidden norm regarding the application of biotechnology to other animals or even humans. Apparently, it is socially acceptable in our culture to manipulate microorganisms and lower animals. This practice does not need any further legitimation.

However, Bacon's text can also be regarded as documenting a 'spirit of optimism' in the light of new possibilities in science and technology. In contrast to Bacon's vision, Kant made a very interesting methodological point more than 100 years later. He shared the euphoria in the field of physics, but was very skeptical about the possibilities in biology. In his *Universal Natural History and Theory of the Heavens* Kant argues:

> In my view, we could say here with certain understanding and without presumption: Give me the material, and I will build a world out of it! That is, give me the material, and I will show you how a world is to come into being out of it. For if there is material present which is endowed with an inherent power of attraction, then it is not difficult to establish those causes which could have led to the arrangement of the planetary system, considered on a large scale. We know what is involved for a body to acquire a spherical shape. We grasp what is required for freely suspended spheres to take on a circular movement around the middle point towards which they are attracted. The position of the orbits relative to each other, the agreement in the direction, the eccentricity, everything can arise from the simplest mechanical causes, and we may hope with confidence to discover them, because they can be established with the easiest and clearest reasons. However, can we boast of such advantages for the smallest plants or insects? Are we in a position to say, give me the material, and I will show you how a caterpillar could have developed? Do we not remain here on the bottom rung because of our ignorance of the true inner constitution of the object and of the development inherent in its multiple elements? (Kant 1968a, pp.229f.)[5]

The demiurgic self-confidence of a typical Newtonian scientist is not transferable to the field of biotechnology. Building the whole immaterial world is countered by the incapability to produce a caterpillar (again a lower animal!). So, in Kant's theory one finds the connection between solid knowledge and knowledge-producing processes as a variant of verumet et factum convertuntur. He sharply separates

---

5   German original: "Mich dünkt, man könne hier in gewissem Verstande ohne Vermessenheit sagen: Gebet mir Materie, ich will eine Welt daraus bauen! Das ist, gebet mir Materie, ich will euch zeigen, wie eine Welt daraus entstehen soll. Denn wenn Materie vorhanden ist, welche mit einer wesentlichen Attractionskraft begabt ist, so ist es nicht schwer diejenigen Ursachen zu bestimmen, die zu der Einrichtung des Weltsystems, im Grossen betrachtet, haben beitragen können. Man weiss, was dazu gehört, dass ein Körper eine kugelrunde Figur erlange, man begreift, was erfordert wird, dass frei schwebende Kugeln eine kreisförmige Bewegung um den Mittelpunkt anstellen, gegen den sie gezogen werden. Die Stellung der Kreise gegeneinander, die Übereinstimmung der Richtung, die Excentricität, alles kann auf die einfachsten mechanischen Ursachen gebracht werden, und man darf mit Zuversicht hoffen sie zu entdecken, weil sie aus die leichtesten und deutlichsten Gründe gesetzt werden können. Kann man aber wohl von den geringsten Pflanzen oder Insect sich solcher Vortheile rühmen? Ist man im Stande zu sagen: Gebt mir Materie, ich will euch zeigen, wie eine Raupe erzeugt werden könne? Bleibe man hier nicht bei dem ersten Schritte aus Unwissenheit der wahren innern Beschaffenheit des Objects und der Verwickelung der in demselben vorhandenen Mannigfaltigkeit stecken?" Translation by Ian Johnston (Kant 2008).

disciplines. The self-confidence witnessed in physics contrasts with a certain humility in biology. Or, as Kant himself states in his *Critique of Judgement*, there will never be a Newton for a blade of grass (cf. Kant 1968b, p.400) Understanding living organisms is strictly decoupled from their technical production. We need other epistemic tools to gain knowledge about living beings.

Kant's rigorous distinction between physics and biology and between technological producibility and teleological interpretation patterns remained valid for a long time. But as a result of developments in biotechnology some philosophers began to diagnose a certain paradigm shift in human self-understanding. Most notably, Hans Jonas and Günther Anders reflected on the change from homo faber to homo creator (cf. Anders 1980, pp.21f.; Jonas 1987, p.30; O. Müller 2012). It is not only about manipulating nature for human purposes and needs, but about inventing new life forms. The human's self-understanding as an inventor and creator changes the sciences.

In the philosophy of science it is commonplace for scientists to not be 'neutral' observers, but involved in the experimental settings. One can now say that not only should the observer's perspective be an issue for meta-reflection on the sciences, but also the implications of different self-understandings. If it is correct to say that the concept of homo creator identifies and addresses a specific profile of self-understanding in the life sciences, then the idea of being creators and the interrelated 'ontological goals' shape the field of research. One symptom might be the way scientists portray themselves. But the design of the experiments is also shaped by the scientists' self-understanding as creators. If the aspect of creation is an essential part of how scientists see themselves, and if the novelty of a life form is a criterion for the quality of research in the life sciences, then one can conclude that the creativity paradigm changes with scientific practice.

One example for the correlation between human self-understanding and changes in scientific practice is the discourse on evolution and the idea that we are reaching the stage where we can take evolution into our hands: "This notion implies not only that humanity is now fully in command of its own destiny, it implies also that we are no longer subject to the haphazard, cumbersome, and often inefficient ways of evolution" (Dyson 2007). A famous quote might prove this demiurgic tendency: "The genetic code is 3.6 billion years old. It's time for a rewrite," said Tom Knight of MIT (cf. Silver 2007). Nordmann draws a link between the attempt at rewriting and the use of the notion of design in synthetic biology:

> This effort consists firstly and primarily of isolating the design efforts from evolutionary processes – be it by studying all organisms as if they were humanly engineered, be it by limiting the work of synthetic biology to closed industrial processes, be it by adopting design constraints that prevent replication, variation, or interaction with

biological systems, or be it by downplaying the likelihood that synthetic biology might alter the course of evolution (Nordmann 2015, p.39).

4 Hybrid practices

Introducing the notion of design is no coincidence. Design appears to be one of the well-defined notions, but it actually reveals a core problem of synthetic biology: the "hybridity" of scientific "practices" that indicates a remarkable blurring of concepts which seems characteristic for synthetic biology (cf. Kastenhofer 2013). In fact, it is unique that we can determine/observe a certain 'inflation' or 'explosion' of terms aimed at capturing what synthetic biologists actually do. When one looks at the impressive list of verbs accumulated over the past years, one can begin to realize what might make synthetic biology intricate. It seems to be difficult to translate what synthetic biologists do into ordinary language. And this is a sign of an ontological, epistemological, and anthropological blurring of boundaries.

The list of activities is not yet complete, but it already includes the following suggestions: modeling life, simulating life, emulating life, modifying life, engineering life, manufacturing life, programming life, designing life, fabricating life, constructing life, creating life, making life, tuning life, tinkering with life, and even kludging life (O'Malley 2011). And one must not forget the epistemologically delicate rewordings: re-modeling life, re-engineering life, re-designing life, re-creating life. Moreover, these courses of action could have non-scientific goals, such as "eliminating the randomness of natural evolutionary advancement," which is an aim of DARPA (Bensaude Vincent 2013a, p.23). These verbal illustrations of self-understanding often include future-driven and techno-utopian ways of scientific practice that typically envision a future with optimally controlled little organisms helping to make the world better.

The above list is remarkable because it is a symptom. It is a symptom of what can be termed 'hybrid practices' which tend to blur the boundaries of scientific methods and other practices such as art and its esthetic norms or economy with their specific infrastructure and organization. My point is that these hybrid practices have their origins in the epistemological black boxes obscuring the exact 'object' of research. Secondly, the human self-understanding as homo creator fosters the ideas of creating and being creative and, consequently, has radicalized scientific practices from knowing to producing.

It is not possible, within the scope of this paper, to delineate all aspects to be considered when discussing hybrid practices. I will only comment briefly on the verb 'design' on the one hand and the idea of 'control' that is implied by the engineering agenda on the other.

Designing seems at first glance to be a scientific method, especially when it refers to 'rational design'. Nordmann analyzes some of the problems encountered with the notion of design:

> How and when is 'to design' the same thing as 'to define' – even if one considers operational rules as definitions, what is defined by the complete 'design' of a minimal system? What makes the synthetic bottom-up design of a minimal cell preferable to the classically analytic 'knock-out' methodology if the aim is to discover the contribution of individual genes to the workings of a cell? And, finally, the editorial states that biology moves beyond the reach of mere intuition when aided by mathematical tools. Does this not imply that knowledge or understanding now reside in the ability to build a computer model, rather than in theories that are tractable by the human mind? (Nordmann 2015, p.49)

Furthermore, the concept of design has non-scientific implications, too. Designing means to shape something in a functional, useful, or even somehow 'beautiful' way. Thus, when designing microorganisms one does not draw exclusively on scientific measurements. Explicitly or implicitly, one is referring to aesthetic norms, too. The notion of 'creation' has similar implications: if it is generally said that scientific practices produce knowledge claims – then how can valid scientific knowledge be identified in design processes? Or, more provocatively, is designing still science? And finally, could the criteria for good design differ from the criteria for good science?

In light of such considerations, one can question synthetic biology's basic hypothesis that producing knowledge and producing things are more or less identical. In the engineering-based configuration of scientific practices, constructing and understanding seem to be the same – in fact, there is a "daunting knowledge gap", as Roberta Kwok (2010) puts it. Some of the epistemological problems regarding the Feynman dictum are mentioned above.

Additionally, I would like to focus on the fact that the engineering agenda is driven by an ideal of 'control'. Several self-descriptions of synthetic biologists allow the conclusion that their main activity is 'controlling' life processes or at least attempting to do so. However, one can ask again: is 'controlling' equivalent to 'knowing' in a scientific sense? Controlling could, of course, be based on knowledge, but a certain hybridity in the scientific practices can again be observed, because controlling implies not only forms of scientific knowledge but also a bio-political dimension. In ethnographic studies synthetic biologists state, for example: "Nature is at our disposal" (Schyfter 2013). This may reflect the good old way scientists have always perceived their work and role, but if Bacon's assumption is still right – that this implies more than undertaking science and producing knowledge, it is about power – one is always operating within a bio-political framework when one tries to extend one's control over the natural domain.

The control and design of organisms as pivotal moments of hybrid practices additionally reveals a number of ethically challenging suppositions. To name only one: controlling and designing organisms appears to be 'pure' scientific practice and, in consequence, seems to be viewed as 'free' research. However, from the observation that synthetic biology and its objectives tend to blur the distinction between scientific and non-scientific criteria and goals one may draw the conclusion that this 'discipline' is a conglomeration of practices that per se transcend the laboratories – and also transcend the laboratories' logical and normative orders and dispositions.

5   Closing remarks: Towards a new theory of values in biotechnology

This perspective leads to ethical implications – not only in terms of adjusting the regularities, but also regarding the 'deep structure' of normative discourses and the societal perception of emerging technologies. In general, one can say that the current profile of synthetic biology and its proponents tends to suggest that the control of life is problematic. Creating life carries the implication of an omnipotent creator who is above all things. Yet, life is possibly not that controllable. The parallels with engines and machines might give rise to an underestimation of life forms that evolve and interact with nature. This in turn could have the effect of underrating the consequences in the long run. Combining the engineerability and evolvability of nature is a thrilling project, but one must beware of the tempting suggestion of life being controllable. Furthermore, the creativity paradigm in the life sciences might not only change scientific practices, but also the value of living beings. The value of a created life form might differ from that of a 'given' life form. As mentioned above, esthetic as well as economic values may come into play when property and copyright issues are involved. For the future debate on synthetic biology it is important to analyze the *hybridity of values* that come together in this domain.

On the basis of an analysis of the multi-faceted dimensions of the hybrid practices in synthetic biology, including the aspects of demiurgic self-assurance, a phenomenology of values connected to the different practices has to be developed. The value of a person or a thing such as a living entity can be revealed by dissecting these practices. The hybridity of practices, which at least includes scientific, esthetic, economic, and bio-political norms, indicates the need to deal with complex phenomena of value conglomerations. By disentangling these conglomerations it will be possible to detect one of the central ethical concerns regarding synthetic biology. The vaguely expressed unease in the media reflects the hybridity of values. And the clues are in the practices: how one deals with living matter says something about its value.[6]

References

Alleyne, R. (2010). Scientist Craig Venter creates life for first time in laboratory sparking debate about 'playing god'. *The Telegraph*, 20 May 2010. http://www.telegraph.co.uk/news/science/7745868/Scientist-Craig-Venter-creates-life-for-first-time-in-laboratory-sparking-debate-about-playing-god.html. Accessed: 16 May 2015.
Anders, G. (1980). *Die Antiquiertheit des Menschen. Über die Zerstörung des Lebens im Zeitalter der dritten industriellen Revolution*. München: Beck.
Bacon, F. (1841). New Atlantis. In: B. Montagu (ed.), *The Works of Francis Bacon, Lord Chancellor of England*, Vol.1 (pp.255-270). Philadelphia: Carey and Hart.
Ball, P. (2004). Synthetic biology: starting from scratch. *Nature, 431*(7009), 624-626. doi: 10.1038/431624a.
Belt, H. v. d. (2009). Playing God in Frankenstein's Footsteps: Synthetic Biology and the Meaning of Life. *Nanoethics, 3*(3), 257-268. doi: 10.1007/s11569-009-0079-6.
Benner, S.A. (2003). Synthetic biology: Act natural. *Nature, 421*(6919), 118. doi: 10.1038/421118a.
Benner, S.A., & Sismour, A.M. (2005). Synthetic biology. *Nature Reviews Genetics, 6*(7), 533-543.
Bensaude Vincent, B. (2013a). Between the possible and the actual: Philosophical perspectives on the design of synthetic organisms. *Futures, 48*, 23-31.
Bensaude Vincent, B. (2013b). Discipline building in synthetic biology. *Studies in History and Philosophy of Biological and Biomedical Sciences, 44*(2), 122-129.
Breithaupt, H. (2006). The engineer's approach to biology. *EMBO Reports, 7*(1), 21-23. doi: 10.1038/sj.embor.7400607.
Brenner, A. (2007). *Leben - Eine philosophische Untersuchung*. Bern: Bundesamt für Bauten und Logistik BBL.
Calvert, J. (2010). Synthetic biology: constructing nature? *The Sociological Review, 58*(s1), 95-112.
Deplazes, A., & Huppenbauer, M. (2009). Synthetic organisms and living machines. *Systems and Synthetic Biology, 3*, 55-63. doi: 10.1007/s11693-009-9029-4.
Dyson, F. (2007). Our biotech future. *The New York Review of Books, 54*(12).
European Commission (2006). *Synbiology. An Analysis of Synthetic Biology Research in Europe and North America*. http://www2.spi.pt/synbiology/documents/news/D11%20-%20Final%20Report.pdf. Accessed: 18 May 2015.

---

6 I thank Ines Anett Schröder for many useful comments to my text.

The Economist (2006). Playing demigods. *The Economist*, 31 August 2006. http://www.economist.com/node/7854771. Accessed: 18 May 2015.

Epping, B. (2008). Synthetische Biologie: Frankensteins Zeit ist gekommen. *Der Spiegel*, 27 December 2008. http://www.spiegel.de/wissenschaft/mensch/synthetische-biologie-frankensteins-zeit-ist-gekommen-a-596579.html. Accessed: 16 May 2015.

Fleck, L. (1979). *The Genesis and Development of a Scientific Fact*. Chicago: University of Chicago Press.

Fox Keller, E. (1995). *Refiguring life: Metaphors of twentieth-century biology*. New York: Columbia University Press.

Fox Keller, E. (2002). *Making Sense of Life: Explaining Biological Development with Models, Metaphors, and Machines*. Cambridge: Harvard University Press.

Gibson, D.G., Glass, J.I., Lartigue, C., Noskov, V.N., Chuang, R.Y., Algire, M.A., ... Venter, J.C. (2010). Creation of a bacterial cell controlled by a chemically synthesized genome. *Science, 329*(5987), 52-56. doi: 10.1126/science.1190719.

Hofmann, M. (2010). Keine Fesseln für die synthetische Biologie. *Neue Zürcher Zeitung*, 11 May 2010. http://www.nzz.ch/aktuell/startseite/keine-fesseln-fuer-die-synthetische-biologie-1.5688716. Accessed: 15 May 2015.

Jonas, H. (1987). *Technik, Medizin und Ethik. Zur Praxis des Prinzips Verantwortung*. Frankfurt/Main: Suhrkamp.

Kant, I. (1968a). Allgemeine Naturgeschichte und Theorie des Himmels oder Versuch von der Verfassung und dem mechanischen Ursprunge des ganzen Weltgebäudes, nach Newtonischen Grundsätzen abgehandelt. In: Kants Werke – Akademie Textausgabe. *Band I: Vorkritische Schriften I (1747-1756)*. Berlin: Walter de Gruyter & Co.

Kant, I. (1968b). Kritik der Urteilskraft. In: Kants Werke – Akademie Textausgabe. *Band V: Kritik der praktischen Vernunft, Kritik der Urteilskraft*. Berlin: Walter de Gruyter & Co.

Kant, I. (2008). Universal Natural History and Theory of the Heavens or An Essay on the Constitution and the Mechanical Origin of the Entire Structure of the Universe Based on Newtonian Principles (trans. by I. Johnston). http://web.calstatela.edu/faculty/kaniol/a360/Kant.Island.Universe.Theory.htm. Accessed: 18 May 2015.

Kastenhofer, K. (2013). Synthetic biology as understanding, control, construction, and creation? Techno-epistemic and socio-political implications of different stances in talking and doing technoscience. *Futures, 48*, 13-22.

Knorr-Cetina, K. (1981). *The manufacture of knowledge: An essay on the constructivist and contextual nature of science*. Oxford, New York: Pergamon Press.

Knorr-Cetina, K. (2009). *Epistemic cultures: How the sciences make knowledge*. Cambridge: Harvard University Press.

Kwok, R. (2010). Five hard truths for synthetic biology. *Nature, 463*(7279), 288-290. doi: 10.1038/463288a.

Latour, B. (1987). *Science in action: How to follow scientists and engineers through society*. Cambridge: Harvard University Press.

Latour, B. (1993). *We have never been modern* (trans. by C. Porter). Cambridge: Harvard University Press.

Luisi, P.L. (2002). Toward the engineering of minimal living cells. *The Anatomical Record, 268*(3), 208-214. doi: 10.1002/ar.10155.

Luisi, P.L., Ferri, F., & Stano, P. (2006). Approaches to semi-synthetic minimal cells: a review. *Naturwissenschaften, 93*(1), 1-13. doi: 10.1007/s00114-005-0056-z.

Müller, H. (2010). Homunkulus muss warten. *Thüringer Allgemeine,* 29 May 2010. http://www.thueringer-allgemeine.de/web/zgt/leben/detail/-/specific/Homunkulus-muss-warten-1148692429. Accessed: 18 May 2015.

Müller, O. (2012). Vom homo faber zum homo creator? Synthetische Biologie und menschliches Selbstverständnis. In: G. Maio (ed.), *Leben schaffen? Philosophische und ethische Reflexionen zur Synthetischen Biologie* (pp.217-230). Paderborn: Mentis.

The New Atlantis (2004). Life from Scratch - Promises, Perils, and Pathogens: Breakthroughs in Synthetic Biology. Editorial. *The New Atlantis,* 5, 101-103.

Nordmann, A. (2015). Synthetic Biology at the Limits of Science. In: B.M. Giese, C. Pade, H. Wigger, A. von Gleich (eds.), *Synthetic Biology: Character and Impact* (pp.31-58): Springer International Publishing Switzerland.

O'Malley, M.A. (2011). Exploration, iterativity and kludging in synthetic biology. *Comptes Rendus Chimie, 14*(4), 406-412.

Panke, S. (2008). Synthetic Biology - Engineering in Biotechnology. Zurich: Committee on applied Bioscience, Swiss Academy of Engineering Sciences. http://www.bsse.ethz.ch/bpl/publications/SATW.pdf. Accessed: 15 May 2015.

Rejeski, D., Schmidt, M., & Wolpe, P.R. (2010). *Transcript of Meeting 1, Session 6 on July 9, 2010.* Washington, D.C.: Presidential Commission for the Study of Bioethical Issues. http://bioethics.gov/sites/default/files/PCSBI%20Meeting%201%2C%20Session%206%20 Transcript.pdf. Accessed: 15 May 2015.

Rheinberger, H.-J. (1997). *Toward a History of Epistemic Things: Synthesizing Proteins in the Test Tube.* Stanford: Stanford University Press.

Schark, M. (2012). Synthetic Biology and the Distinction between Organisms and Machines. *Enviromental Values, Vol. 21,* 19-41. doi: 10.3197/096327112X13225063227943.

Schmidt, M. (2009). Do I understand what I can create? Biosafety issues in synthetic biology. In: M. Schmidt, A. Kelle, A. Ganguli, H. de Vriend (eds.), *Synthetic biology. The technoscience and its societal consequences* (pp.81-100). Dordrecht: Springer Academic Publishing.

Schyfter, P. (2011). Technological biology? Things and kinds in synthetic biology. *Biology and Philosophy, 27*(1), 29-48. doi: 10.1007/s10539-011-9288-9.

Schyfter, P. (2013). How a 'drive to make' shapes synthetic biology. *Studies in History and Philosophy of Science Part C: Studies in History and Philosophy of Biological and Biomedical Sciences, 44*(4), 632-640.

Silver, L. (2007). Scientists Push the Boundaries of Human Life. *Newsweek,* 3 June 2007. http://www.newsweek.com/scientists-push-boundaries-human-life-101723. Accessed: 15 May 2015.

Der Spiegel (2010). Durchbruch: Forscher erschaffen erstmals künstliches Leben. *Der Spiegel,* 20 May 2010. Retrieved from http://www.spiegel.de/wissenschaft/natur/durchbruch-forscher-erschaffen-erstmals-kuenstliches-leben-a-696016.html. Accessed: 15 May 2015.

Tucker, J.B., & Zilinskas, R.A. (2006). The promise and perils of synthetic biology. *The New Atlantis, 12,* 25-45.

Living Machines
On the Genesis and Systematic Implications of a Leading Metaphor of Synthetic Biology

Harald Matern, Jens Ried, Matthias Braun and Peter Dabrock

The current developments and (postulated) products in the field of synthetic biology have given rise to numerous metaphors within the discourses on interpretation and plausibility. Apart from the metaphors seeming to be especially 'prominent', such as 'playing God' or 'creating life' (Ried and Dabrock 2011), another 'player' appears on the field of strategies of explanation and functionalization: 'living machines'. Thus Tucker and Zilinskas (2006) say: "Synthetic biology is another transformative innovation that will make it possible to build living machines from off-the-shelf chemical ingredients [...]". If 'living machines' are mentioned in the current debate on synthetic biology, this term aims at describing entities which are commonly described as 'machines' due to their genesis and functionality on the one hand, but which can simultaneously be labeled as 'living' or as organisms due to their composition of biological items and their future functioning on the other hand. This definition differs from the use of the term in the recently emerging field of research which is known as "biomimetics" (cf. Prescott et al. 2012, Lepora et al. 2013). While Synthetic Biology points at constructing life from scratch, biomimetics focuses "on technologies that emulate living organisms – living machines" (Prescott et al. 2014). The following aims at a better understanding of the genesis and systematic implications of "living machines" as a leading metaphor of synthetic biology.

Whereas at a first glance *live* or *living* on the one hand and *machine* on the other hand seem to be incompatible epistemological and ontological descriptions of the very same entity, the metaphor 'living machines' hints at the possibility of such an inverted intertwining of both terms: Whereas organisms are traditionally unambiguously assigned to the realm of the living, machines are most of all regarded in the focus of functionality, the predictable and non-autotelic creation and preservation. The linguistic ambiguity of the 'living machines' shown becomes even more complex when perceiving that the term 'living machines' is no vanguard of the current debates but can genealogically already be found in classical philosophical deliberations. Thus the intentional content of the metaphor shall be more closely explored and determined in the following reflections by seeking out the genealogy of the metaphor 'living machines' at its historical place as a first step. Here it will be determined by disputing with Immanuel Kant in how far the unsolvable ambiguity of the relationship between organism and machine can be tried to be understood as an 'abundance' of life. In a second step the results thus gained will be modeled due to their systematical-ethical consequences.

1 Historical Lines

Immanuel Kant developed his up to now influential[1] conception of life on the background of dealing with empirism and rationalism. Here Kant's critique of reason focuses on the *epistemological* problem which is reflected in his deliberations on the organic in such a way that it is standing at a double position between "nature" and "freedom" (*ontological*) on the one hand and between phenomenality and noumenality (*epistemological*) on the other hand. The identified tension leads Kant to using a compound metaphor for the phrasing of this problem in order to blend both sides.[2] Kant resorts to the machine imagery (cf. Jahn 2002, pp.196-212; Gloy 1995,

---

1   Still today Kant's deliberations on the organism are seen as a challenge to the methodology of strategies of description that are arguing strictly 'empirically', cf. Moya et al. (2009, p.28): "The history of biological research can be regarded as an attempt to prove Kant wrong."
2   By doing so, Kant already undermines the differentiation of "explaining" and "understanding" used since Dilthey for the distinction of "natural sciences" and "humane sciences". Kant uses the term "explanation" only to refer to processes and "appearances" conceivable by natural laws (for Kant: mainly by causality), cf. "In der Erklärung der Erscheinungen der Natur muß uns indessen vieles ungewiß und manche Frage unauflöslich bleiben, weil das, was wir von der Natur wissen, zu dem, was wir erklären sollen, bei weitem nicht in allen Fällen zureichend ist" (Kant 1787, p.505). "Die Erscheinungen verlangen nur erklärt zu werden, so weit ihre Erklärungsbedingungen in der Wahrnehmung gegeben sind, alles aber, was jemals an ihnen gegeben werden mag, in einem absoluten Ganzen zusammengenommen, ist selbst eine Wahrnehmung. Dieses All aber ist es eigentlich, dessen Erklärung in den transzendentalen Vernunftaufgaben gefordert wird" (Kant 1787, pp.511f.). "Wir können nichts erklären, als was wir auf Gesetze zurückführen können, deren Gegenstand in irgend einer möglichen Erfahrung gegeben werden kann." (Kant 1785, p.459). In his *Critique of Judgment* Kant contraposes mechanism and teleology as two different "Erklärungsart[en]" (Kant 1793, pp.355,357 a.m.). The "explanation" always follows "gegebenen Gesetzen" – "[d]enn Erklärung heißt von einem Prinzip ableiten, welches man also deutlich muss erkennen und angeben können" (Kant 1793, p.358).

pp.162-172) prominently used especially in the context of the Enlightenment and the so-called French Materialism[3] for the interpretation of "nature", but he adds the epithet "living". It refers to the central question of the integrating principle of the mechanistically understood body, which induces the self-referential characteristics of the organism. Kant argues – in contrast to La Mettrie – on the background of his epistemological deliberations that such a principle has to be assumed as necessary but immaterial. The question of the 'life' of the living, or in other words, the question of the principle of the unity and identity of the organism becomes apparent as a problem in Kant's position – and the associated conventions of speech become clear as descriptions of a systematic shifting of boundaries. Whereas the semantic focus previously lay on the metaphor 'machine', now it lies on Kant's expression "living".

What is relevant about Kant's considerations for the current debate on the term and phenomenon of life is the circumstance that Kant's considerations consist of the drawing of lines between the exactly knowable and the only imaginable objects of knowledge and at the same time the thus related ontological lack of clarity in the context of the synthetic biology that is actualized and potentiated at exactly that point at which "life" shall be artificially *created*. This 'grey area' of epistemology and ontology which is caused by this blurring of boundaries is characterized by the metaphor of the 'living machine' in both cases.

The term explanation was later on taken up by Wilhelm Dilthey in order to distinguish the specific recognitional way of the "natural sciences" from those of the "humane sciences" (the understanding) (cf. Dilthey 1883, pp.81-84,86-88,117-120). As opposed to the "understanding" of humane sciences, the natural sciences use the method of explanation due to "hypothetischen Annahmen" (Dilthey 1883, p.118), as they ascribe a regular relationship to the exactly knowable individual phenomena "physische[n] Tatsache[n]" (Dilthey 1883, p.86): "[D]amit ist nun der Unterschied beider Arten von Wissenschaften gegeben. In der äußeren Natur wird Zusammenhang in einer Verbindung abstrakter Begriffe den Erscheinungen unterlegt. Dagegen der Zusammenhang in der geistigen Welt wird erlebt und nachverstanden." (Dilthey 1883, p.119) – "Dort werden für die Individuation hypothetische Erklärungsgründe aufgesucht, hier dagegen werden in der Lebendigkeit die Ursachen derselben erfahren" (Dilthey 1883, p.120).

3   Not only would Julien de La Mettrie, who transferred the mechanistic world view to living bodies in the wake of Descartes (cf. Gloy 1995, pp.165f.) and thus assumed a material inherent principle of motion (The human body "ist eine Maschine, die selbst ihre Triebfedern aufzieht" (La Mettrie 2009, p.35)), or Paul Henri Thiry d'Holbach, according to whom life is "nur eine Vereinigung von Bewegungen" (d'Holbach 1841, p.28), have to pay attention to a more thorough description but additionally they would have to consider Maupertuis and Denis Diderot, as well as especially the materialism of the 17th century of the English-speaking world. Cf. Panajotis Kondylis (1986, pp.257-290) for the origin and distribution of the materialism in the Age of Enlightenment.

According to Kant *life* is the "Vermögen einer Substanz, sich aus einem inneren Princip zum Handeln, [...] zur Veränderung [...] zu bestimmen" (Kant 1781, p.544). This expression points at the metaphysical-critical deliberations already made in the *Critique of Pure Reason* taking the example of the contrast of "freedom" and "nature".[4] A similar problem arises when trying to conceptualize 'life'. Thus life is a form of causality which itself cannot simply be explained in an empirical-causal way.

Apart from defining life as the ability of self-determination, Kant develops a second aspect which is related to his elaborations on the *organism*. The starting point for the question thus is the "objektive Zweckmäßigkeit der Natur" (Kant 1793, p.267)[5] assumed as due to knowledge by Kant in his *Critique of Judgment*. This means that when attempting to describe relations in nature, obviously not only causal-mechanical assumptions are predominant but also such teleological patterns of explanation[6] "die eine besondere Art der Kausalität, nämlich die Zwecke (*nexus finalis*) zu Hilfe nehmen" (Kant 1793, p.269). Here Kant presupposes that the "Zwecke" are cognitive requirements necessary for explanation. However, Kant states that this "transzendentale Begriff einer Zweckmäßigkeit der Natur" would be neither "ein Naturbegriff noch ein Freiheitsbegriff, weil er gar nichts dem Objekte (der Natur) beilegt" (Kant 1793, p.XXXIV). Nonetheless, directly afterwards he points out that "wir auch, gleich als ob es ein glücklicher, unsere Absicht begünstigender Zufall wäre, erfreut (eigentlich eines Bedürfnisses entledigt) werden, wenn wir eine solche systematische Einheit unter bloß empirischen Gesetzen antreffen" (Kant 1793, p.XXXIV).

What is blurred due to this "fortunate coincidence" is the differentiation, which is otherwise sharply made by Kant, between phenomenal and noumenal knowledge – or in this case between epistemology and ontology. This fundamental blurring of boundaries can be also seen for the "innere Zweckmäßigkeit des Naturwesens" (Kant 1793, p.280) assumed by Kant, if transferred to an individual object of knowledge. It is characterized by inherently postulating the principle of the unity of the parts of a natural being. Thus, according to Kant, the natural purpose demands

---

4   Accordingly Kant states in the resolution of the third antinomy in the "transcendental dialectics" of his "Critique of Pure Reason": "In Ansehung dieses empirischen Charakters [cf. of each action in so far as it is object of the observation] gibt es also keine Freiheit, und nach diesem können wir doch allein den Menschen betrachten, wenn wir lediglich beobachten, und, wie es in der Anthropologie geschieht, von seinen Handlungen die bewegenden Ursachen physiologisch erforschen wollen" (Kant 1787, p.578).

5   Cf. also Neugebauer (2010, pp.26ff.) for nature's purpose as a "leading category" of Kant's conception of life.

6   Kant understands "teleology" as way of explaining phenomena contrary to the mechanist view, cf. Kant (1793, pp.355f.).

an explanation considering the "Kausalität der Begriffe von vernünftigen Wesen außer ihm" (Kant 1793, pp.290f). A natural product can consequently be considered a natural purpose "wenn es von sich selbst [...] Ursache und Wirkung ist" (Kant 1793, p.286, highlighted in the original). The being determined as the natural purpose has to be understood as an "organized being", according to Kant. As such, it is simultaneously an "Analogon des Lebens" (Kant 1793, p.293)[7] – though Kant himself questions the use of the term "analogy".[8] The expression of the analogy of 'life' is further determined as deduction "nach einer entfernten Analogie mit unserer Kausalität nach Zwecken überhaupt" (Kant 1793, p.295)[9]. From this it can be concluded that according to Kant 'life' is not realized by the empirical characteristics of the *living* – and it resists definite rational knowledge.

The organism as a whole determined as such (i.e. as a natural purpose) possesses the ability of self-organization and growth.

> Ein organisiertes Wesen ist also nicht bloß Maschine, denn die hat lediglich bewegende Kraft, sondern es besitzt in sich bildende Kraft, und zwar eine solche, die es den Materien mitteilt, welche sie nicht haben (sie organisiert), also eine sich fortpflanzende bildende Kraft, welche durch das Bewegungsvermögen allein (den Mechanism) nicht erklärt werden kann (Kant 1793, pp.292f).

Thus, what remains for the knowledge is an undissolvable 'surplus' for the object which differentiates it from the machine, as Kant explains when referring to the

---

7 Such a virtually analogous concept of life has already been created before Kant. In another context it can be found in Leibniz' works. Here it is directed at the analogy of living organisms and the metaphysical ground of the unity of the world. Despite the difference of the epistemological basic assumptions between Leibniz and Kant, Leibniz' analogous way of regarding the organism is a preliminary stage to Kant's subjectivity-theoretical constructivism, according to Cheung (2000, pp.51ff.).
8 Kant 1793, 293: "Genau zu reden hat also die Organisation der Natur nichts Analogisches mit irgendeiner Kausalität, die wir kennen."
9 Angela Breitenbach stresses the constructivist character of such an analogous deduction: "Erst indem wir die Fähigkeit unseres eigenen Vernunftvermögens, Zwecke zu setzen und nach diesen Zwecken zu streben, in die Natur hineinlesen, können wir die Natur Kant zufolge auch als lebendig betrachten" (Breitenbach 2009, p.31). However it does not suffice to explain the process by saying "dass die menschliche Vernunft in der Betrachtung eines beliebigen belebten Wesens – sich selbst begegnet" (thus Gerhardt 2002, p.317, quoted in Breitenbach 2009, p.32). This would presuppose a clear concept of reason and nature. In contrast to this, the question of life, as asked by Kant, points at the fact that the lines between nature and reason are blurring here.

metaphor of the clockwork.¹⁰ At the same time Kant vehemently fights against regarding this 'surplus' as positive.¹¹ In his later works Kant develops the expression of the organisms as 'living machines'.¹² The 'surplus' of the self-organizing living regarding the mechanic conception of the living organism is marked in the semantics of the epitheton.

As a conclusion it can be said: The term 'living machines' in this context turns out to be a metaphor that firstly, in a historical-systematical way, refers to Kant's account of purely mechanically argumentative 'explanations' of the organic and that secondly describes the double position of the organic as a matter of phenomenal and noumenal knowledge. Thirdly the use of the metaphor points out that 'living'

---

10 Cf. Kant (1793, pp.292f.): "Daher bringt auch so wenig, wie ein Rad in der Uhr das andere, noch weniger eine Uhr andere Uhren hervor, so daß sie andere Materie dazu benutzte (sie organisierte); daher ersetzt sie auch nicht von selbst die ihr entwandten Teile, oder vergütet ihren Mangel in der ersten Bildung durch den Beitritt der übrigen, oder bessert sich etwa selbst aus, wenn sie in Unordnung geraten ist: welches alles wir dagegen von der organisierten Natur erwarten können. – Ein organisiertes Wesen ist also nicht bloß Maschine [...]."

11 Just cf. the expressions of the introduction of the *Critique of Judgment* mentioned above (Kant 1793). At another point, Kant also speaks of a "übersinnliche[s] Substrat der Natur", "von dem wir nichts erkennen", and of the circumstance that the relational determination of mechanic explanation and teleological assessment of nature is "für unsere Vernunft auf immer unbestimmbar" and eventually leads to a purely "intelligibelen Prinzips der Möglichkeit einer Natur überhaupt" (Kant 1793, p.362). Cf. also the evidence quoted by Heimsoeth (1940, pp.95-98) from the posthumous work of Kant's 'immateriality' of this principle.
Such an affirmation of the 'surplus' can be found in the earlier works of Schelling. Schelling assumes that Kant's 'surplus' – which entails within the power as a regulating principle of the knowledge of 'life', analogously to freedom presupposed by epistemological reasons, the presupposition of an intrinsic principle of the unity of the organism as the principle of life due to term logic – could be described as the constitutive principle of the entire natural context, as world soul (cf. Schelling 1798, VI., p.226).

12 Cf. the evidence quoted by Heimsoeth from the posthumous work (Heimsoeth 1940, pp.89f. w. annotation 25): "natürliche Maschinen"; bodies "welche *aus eigenen Kräften* als Maschinen wirken"; "sich selbst der Form nach erzeugende Maschinen"; "Körper, deren innere Formen als absichtlich, d.i. nach einem Princip der Zwecke möglich gedacht werden können, sind organische [...] und sind Maschinen, entweder *leblose* (blos vegetierend) oder *animalisch* lebende"; (ibid., p.94 w. annotation 50): "Der organische Körper ist also Maschine, deren bewegende Kraft absolute Einheit der Verbindung der durch jenes Subject bewegten Materie bey sich führt."
The still tentative and experimental use not only of the metaphor of 'living machines' but similar metaphors points at the vagueness inherent to the subject of 'life' – respectively the crossing of borders due to the constitution of the subject.

is an attribute attached to the organism (which is itself 'understood' as a machine) out of embarrassment.[13] Thus it follows:

Firstly, it is the internal structure that is fundamental for the differentiation of organism and machine.[14] Secondly, the organism is partly a product of itself. This is referred to as 'self-organization'.[15] However, the machine is a pure product of human *techne*. Thirdly, it follows that the purpose[16] of machines is exterior whereas the purpose of organisms has to be at least considered as intrinsic. Fourthly, machines can thus be explained just by causal laws according to their manner of functioning and their genesis whereas a further perspective of explanation has to be added when regarding organisms, which is the perspective of teleology for Kant.

In order to determine the relationship between organism and machine, two areas of questions have to be formulated: Firstly, it has to be determined in how far the notion 'living machines' can present the line between organism and machine, or respectively between life and machine in no way as a rigid, topographically locatable line of distinction, but shows such a line as moveable dependent on (bio-) technological developments. Thus the metaphor of the 'living machines' could be understood as a point of condensation where the shifting of boundaries, the eruption of the previous distinctions, becomes observable and processable.[17]

When attempting to mark the drawing of boundaries or the shifting of boundaries, it secondly has to be examined in how far such a drawing of boundaries can be made by a closer determination of the 'surplus'. Following this line of examination, with the help of the figure of 'surplus' it would be expressed that there is always at least one dimension of the living which we cannot comprehend. Such a statement on the withdrawal of the living can only be understood correctly if it is understood as the impossibility of *simultaneously* taking up all perspectives necessary for comprehending life. Such an understanding of the 'surplus' states that a determination of that which we denote as living is not simply solved into a leveling intertwining

---

13 Cf. similarly Alt (2010, p.164). It does not represent a further development of the concept of life to limit this concept to human life due to lack of knowledge of the theological perspective, as Neugebauer (2010, pp.335ff.) does.
14 Cf. also: Müller-Strahl 2011.
15 The expression 'self-organization' is currently still a major metaphor when discussing the adequate description of the living, cf. Johnston (2008, pp.188-195).
16 When talking about purpose, this does not have to be an empirically verifiable purpose. In the present context the concept of the purpose is a presupposition necessary due to epistemological reasons.
17 Thus Johnston (2008, p.165) states: "The very idea of self-reproducing machines disrupts our conceptual boundaries"; cf. pp.166f. referring to Kant. Therefore, 'life' would have to be attributed to the self-organizing machine: cf. p.167.

of the compounds of *life* and *machine* by talking about 'living machines'. Instead, both poles still have a tensed relationship in the context of metaphorical speech. At the same time it has to be cautioned against misunderstanding such a 'surplus' of the living compared to the machine as vitalistic without considering its interpretation-theoretical and term-logical implications. Both these questions shall be dealt with in the concluding part.

2 Systematic-Ethical Explorations

The metaphor of the 'living machines' (Deplazes and Huppenbauer 2009; Potthast 2009) is in the eye of the public perception *and* of scientific descriptive heuristics. Already at an earlier point in the development of 'synthetic biology', this metaphor was used to denote the 'new' connection and genuine integration of engineering into biology. Likewise, one of the pioneers of synthetic biology, Steve Benner, already very early talked about (the necessity of) a mechanization of biology and an opportunity thus resulting: "redesigning life" (Ball 2004). Drew Endy takes up this term and uses it in order to outline the possibilities of success of the synthetic biology which are exactly arising from the fact that direct parallels can be drawn between the 'living machines' and other machines like computers, mobile phones or cars (Endy 2005; Endy 2008). Similarly, 'living machines' describe the basic idea as well as the purpose of the iGEM competition (Mitchell et al. 2010; Dixon and Kuldell 2011).

However, metaphors must not only be understood as figures of speech but as forms of speech that express the exterior of those metaphorical concepts which guide and pervade our interpretation and formation of reality (Lakoff and Johnson 2003, p.7) and which, as "speech acts" (Austin 2002), do not only have communicative and interpretative functions, but also structuring, orienting and active guiding functions (Lakoff and Johnson 2003, pp.14-21,61-68).

Considering the expected innovations in this field, the metaphor 'living machines' does not only denote the attempt to find forms of speech which make communicable the characteristics of the scientific field and the targeted products, but it is also an indicator of the uncertainty which synthetic biology creates by propagating the crossing of the classical distinctions ('life' vs. 'non-life'; 'natural' vs. 'artificial') and which first and foremost makes the search for an adequate linguistic representation necessary and propels it. However, as necessary as the use of metaphors is for a linguistic interpretation of new fields of reality, as problematic can be the further associations invoked by the metaphors and their backgrounds.

The historical analysis of the term 'living machines' shows that – explicitly in contrast to the leading intentions which stimulate their use in the current discourse, indeed – this metaphor links forms of speaking and forms of thinking in a way that invites a vitalistic interpretation. In order not to commit such a fallacy concerning conceptual history by drawing conclusions for topical questions just from the historical genesis, it (1) remains to be examined if these ambiguities are of a systematic nature and are inherent to the concept. Subsequently, it has to be examined (2) whether such a tension in the expression 'living machines' can be solved, e.g. in such a way that this 'vacancy' is reinterpreted in a normative way – or respectively as the 'surplus' at which it hints.

The metaphor 'living machines' is characterized by linking two realms that are mostly used separately in everyday communication. The perspectives metaphorically linked by 'living machines' are characterized by being neither reducible to each other nor can they communicate separately what is considered to be the specific of the (actual and expected) products of synthetic biology. Thus it is the constitutive tension which is thus not solvable or the complementarity of the perspectives that only result in a complete description by their mutually exclusive and yet postulated unity. Inasmuch that the complementary tension between the aspects of 'life' and 'machine' is fundamental for the metaphor examined here, what has to be stated while principally agreeing with the historical considerations made above is that there is an unsolvable 'open spot' also in a systematic way; and this spot is functionally identical with the 'surplus' conceptualized as specific for the 'living machines' by Kant. On the one hand this 'surplus', or respectively this 'blank space', marks a gateway for fillings of differing provenance and on the other hand it gives rise to the question of its further processing in the discourse on synthetic biology.

The fact that in the case of the 'living machines' filling the blank space in the sense of vitalism was historically and systematically common and can currently not be excluded (Deplazes and Huppenbauer 2009) points at a decline within the metaphor from the aspect of 'life' to the aspect of 'machine'. However, at the same time stating such a decline evokes the danger of affirming the surplus element of life and thus of eliminating the constitutive ambiguity of the 'living machines'. Rightly, it can be argued with Kant and even beyond Kant that each attempt of positively affirming – be it of a vitalistic or another nature – eventually contradicts the communicative function of this metaphor. The 'living machines' do win their plausibility within the scientific and public discourses on synthetic biology exactly by sustaining the tension between both compound elements: 'life' and 'machine'.

If regarding both compounds more precisely, it becomes apparent that in contrast to the common laws of compound building the conceptual weight does not lie on the second but actually on the first element. Coming into focus, the demand to more thoroughly look at the aspect 'living' and to analyze it is in line with interest in the historical-conceptual (re-)interpretation of the concept 'life', which is growing in parallel to the development of synthetic biology.

The challenge was conceptualized in a historical perspective as 'surplus' of life – while this 'more' appears as itself regarding the aspect of the machine and not the aspect of life which represents this 'surplus' in a way that has yet to be specified – and in a systematic perspective as opening gap due to the unsolvable because conceptually-constitutively complementary tension between the association fields of 'life' and machine. In the discourse on synthetic biology this challenge can and will be among other things assessed as a normative claim that life has for us and that distinguishes life from the purely material (Boldt et al. 2009). This tendency does not only reconnect the above-mentioned decline within the metaphor with the primacy of life, but it also deduces the central meaning of the concept of life as the focal point of differing questions and criticism, especially in the normatively oriented debates on synthetic biology (Dabrock et al. 2011).

It seems to be doubtful whether the ethical challenge that the question of the "mechanization of the living" entails can be reasonably faced with the assumption of a special normative demand of the living in contrast to the merely material (Boldt and Müller 2008; Boldt et al. 2009), especially as the problem of the 'blank space' brought up by the metaphor is only moved from the terminological-conceptual level to an epistemological level – condensed in the question of the possible metaphysical backgrounds and undergrounds of this special claim. Furthermore, what is connected to such a figure of thought which claims a special normative demand of life is the postulate of a definitional unambiguity of that which is 'life', which de facto cannot be fulfilled.

The ethical challenges transported in the metaphoric content of the 'living machines' can thus neither lie in a one-sided focus on questions of the 'vitalization of the machine' nor in the one-sided focus on the 'mechanization of the living', but rather in the observation and precise exploration of the ambiguity and contextual shifting in the relationship of the organism and the machine.

On the contrary, ethics now faces the task of determining boundaries in the relationship of the organism and the machine, conceptually linked to the developments in the field of biotechnology. Such a contextual determination of boundaries also has to be questioned regarding its relation to the self-concept of scientists.

Thus it can be stated for the field of synthetic biology that the metaphor of the 'living machines' is not only a central metaphor in the public perception of synthetic biology (Marlière et al. 2011), but it is also used by the researchers involved (Kamm and Bashir 2014). The self-description of the scientists (as those who produce 'living machines') thus entails a moment of vagueness or ambiguity in a central aspect.[18] On the one hand this vagueness points at the undermining constitutive for synthetic biology of the differentiation of basic and applied science which is otherwise common to scientific research. On the other hand especially the choice of the metaphor for denoting the object of research or the product created is a statement of awareness of the scientists concerning the respective object. It is worthwhile to examine whether the choice of the metaphor 'living machines' by the scientists does – apart from the undeniable pursuit of affecting the media and the public – not also express an uncertainty on the right placing of the object researched and produced in the chiastic relationship of the mechanization of the living and the vitalization of the machine. Thus, the genius of the self-concept of the scientists could exactly lie in the fact that they use the metaphor of the 'living machines' to describe that – according to Nietzsche – it not only comes to a "grosse[n] Kluft zwischen dem Künstler, der auf abgelegener Höhe sein Werk schafft, und dem Publicum, welches nicht mehr zu jener Höhe hinaufkam und endlich missmutig wieder tiefer hinabsteigt" (Nietzsche 1980, p.168), but that this gap, understood as ambiguous fascination as well as tentative astonishment, is also building up between the scientists and their object, which is tried to be described by the metaphor 'living machines'. For ethical research, the dealing with the ambiguity of this metaphor allows on the one hand to try to get to the bottom of the determination of boundaries between a mechanization of life and a vitalization of the machine, on the other hand it allows to examine the relation of such an ethical determination of boundaries to the self-concept of the researchers in synthetic biology, i.e. the possibility to focus on the self-concept of the researchers as a fundamental marker of novelty of synthetic biology (Schummer 2011; Dabrock et al. 2011, p.393) and to inquire for consequences of the ethical research in the field of synthetic biology.

---

18 On the one hand this vagueness makes it possible to juggle with the relationship of the mechanization of the living and the vitalization of the machine; however, it eventually remains undeterminable. According to Kant, it can be stated that even deciding for a pure mechanization concept does eventually not enable the undermining of this ambiguity.

References

Alt, W. (2010): Systemtheoretische Prinzipien des Lebendigen – Emergenz von "Funktionen" und "Spielräumen". In: V. Herzog (ed.), *Lebensentstehung und künstliches Leben. Naturwissenschaftliche, philosophische und theologische Aspekte der Zellevolution* (pp.160-213). Zug/Schweiz: Die Graue Edition.

Austin, J.L. (2002). *Zur Theorie der Sprechakte (How to do things with Words)*. Stuttgart: Reclam (Original: 1962).

Ball, P. (2004). Synthetic biology: starting from scratch. *Nature, 431,* 624-626.

Becker, C. (2009). Einleitung. In: J.O. de La Mettrie, *Die Maschine Mensch. Französisch – Deutsch. Übersetzt und herausgegeben von Claudia Becker* (pp.VII-XX). Hamburg: Meiner.

Boldt, J., & Müller, O. (2008). Newtons of the leaves of grass. *Nature Biotechnology, 26,* 387-389.

Boldt, J., Müller, O., & Maio, G. (2009). *Synthetische Biologie. Eine ethisch-philosophische Analyse.* Bern: BBL.

Breitenbach, A. (2009). Die Frage nach dem Lebendigen in Zeiten biowissenschaftlichen Fortschritts. In: S. Springmann, A. Trautsch (eds.), *Was ist Leben? Festgabe für Volker Gerhardt zum 65. Geburtstag* (pp.29-34). Berlin: Duncker & Humblot.

Cheung, T. (2000). *Die Organisation des Lebendigen. Die Entstehung des biologischen Organismusbegriffs bei Cuvier, Leibniz und Kant.* Frankfurt, New York: Campus.

Dabrock, P., Bölker, M., Braun, M., & Ried, J. (eds.) (2011). *Was ist Leben – im Zeitalter seiner technischen Machbarkeit?* Freiburg: Alber.

Deplazes, A., & Huppenbauer, M. (2009). Synthetic organisms and living machines. Positioning the products of synthetic biology at the borderline between living and non-living matter. *Systems and Synthetic Biology, 3*(1-4), 55-63.

Dilthey, W. (1883). *Der Aufbau der geschichtlichen Welt in den Geisteswissenschaften.* Gesammelte Schriften Vol. 7 (1979). Göttingen: Vandenhoeck und Ruprecht.

Dixon, J., & Kuldell, N. (2011). BioBuilding: Using Banana-Scented Bacteria to Teach Synthetic Biology. *Methods in Enzymology, 497,* 255-271.

Endy, D. (2005). Foundations for engineering biology. *Nature, 438,* 449-453.

Endy, D. (2008): *Drew Endy on Engineering Biology.* Soft Machines. http://www.softmachines.org/wordpress/?p=389. Accessed: January 2012.

Gerhardt, V. (2002). *Immanuel Kant. Vernunft und Leben.* Stuttgart: Reclam.

Gloy, K. (1995). *Das Verständnis der Natur. Vol. 1: Die Geschichte des wissenschaftlichen Denkens.* München: C.H. Beck.

Gunderson, K. (1964). Descartes, La Mettrie, Language and Machines. *Philosophy, 39*(149), 193-222.

Heimsoeth, H. (1940). Kants Philosophie des Organischen in den letzten Systementwürfen. Untersuchungen aus Anlaß der vollendeten Herausgabe des Opus postumum. *Blätter für deutsche Philosophie, 14* (1940/41), 81-108.

d'Holbach, P.H.T. (1841). *System der Natur von Mirabaud. Deutsch bearbeitet und mit Anmerkungen versehen.* Leipzig: Georg Wigands Verlag.

Jahn, I. (2002). Naturphilosophie und Empirie in der Frühaufklärung (17. Jh.). In: I. Jahn (ed.), Geschichte der Biologie (pp.196-212). Heidelberg, Berlin: Spektrum.

Johnston, J. (2008). *The Allure of Machinic Life: Cybernetics, Artificial Life, and the New AI.* Cambridge, London: MIT Press.

Kamm, R.D., & Bashir, R. (2014). Creating Living Cellular Machines. *Annals of Biomedical Engineering 42*(2), 445-459.

Kant, I. (1781). Grundlegung zur Metaphysik der Sitten. In: Kants Werke – Akademie Textausgabe. *Band IV: Kritik der reinen Vernunft (1. Aufl. 1781), Prolegomena, Grundlegung zur Metaphysik der Sitten, Metaphysische Anfangsgründe der Naturwissenschaften* (1968). Berlin: Walter de Gruyter & Co.

Kant, I. (1786). Metaphysische Anfangsgründe der Naturwissenschaften. In: Kants Werke – Akademie Textausgabe. *Band IV: Kritik der reinen Vernunft (1. Aufl. 1781), Prolegomena, Grundlegung zur Metaphysik der Sitten, Metaphysische Anfangsgründe der Naturwissenschaften* (1968). Berlin: Walter de Gruyter & Co.

Kant, I. (1787). *Kritik der reinen Vernunft*. Zweite Auflage. Riga.

Kant, I. (1793). *Kritik der Urtheilskraft*. Berlin.

Kondylis, P. (1986). *Die Aufklärung im Rahmen des neuzeitlichen Rationalismus*. München: dtv.

La Mettrie, J.O. d. (2009). *Die Maschine Mensch. Französisch – Deutsch. Übersetzt und herausgegeben von Claudia Becker*. Hamburg: Meiner.

Lakoff, I., & Johnson, M. (2003). *Metaphors we live by*. London, Chicago: University of Chicago Press.

Lepora, N., Mura, A., Krapp, H.G., Verschure, Paul. F.M.J., & Prescott, T.J. (2013). *Biomimetic and Biohybrid Systems*. Second International Conference, Living Machines 2013, London, UK, July 29 – August 2, 2013. Proceedings. Berlin, Heidelberg: Springer.

Marlière, P., Patrouix, J., Döring, V., Herdewijn, P., Tricot, S., Cruveiller, S., . . . Mutzel, R. (2011). Chemical evolution of a bacterium's genome. *Angewandte Chemie International Edition, 50*(31), 7109-7114.

Mitchell, R., Dori, Y.J., & Kuldell, N. (2010). Experiential engineering through iGEM— An undergraduate summer competition in synthetic biology. *Journal of Science Education and Technology, 20*, 156-160.

Müller-Strahl, G. (2011). Metaphysik des Organismus und die Erklärung organischer Formen des Lebendigen. In: P. Dabrock, M. Bölker, M. Braun, J. Ried (eds.), *Was ist Leben – im Zeitalter seiner technischen Machbarkeit?* (pp.195-226). Freiburg: Alber.

Neugebauer, M. (2010). *Konzepte des "Bios". Leben im Spannungsfeld von Organismus, Metaphysik, Molekularbiologie und Theologie*. Göttingen: Ed. Ruprecht.

Nietzsche, F. (1980). *Menschliches, Allzumenschliches I und II*. Kritische Studienausgabe Band 2 (KSA2). Berlin, New York: de Gruyter.

Potthast, T. (2009). Paradigm shifts vs. fashion shifts? Systems and synthetic biology as new epistemic entities in understanding and making "life". *EMBO Reports, 10* (Special Issue), 51-54.

Prescott, T.J., Lepora, N.F., Mura, A., & Verschure, P.F.M.J. (2012). *Living Machines: The first international conference on biomimetic and biohybrid systems*. London: Springer.

Prescott, T., Lepora, N., & Vershure, P.F.M.J. (2014). A future of living machines? International trends and prospects in biomimetic and biohybrid systems. *Proc. SPIE 9055, Bioinspiration, Biomimetics, and Bioreplication 2014*, 905502. doi:10.1117/12.2046305.

Ried, J., & Dabrock, P. (2011). Weder Schöpfer noch Plagiator. Theologisch-ethische Überlegungen zur Synthetischen Biologie zwischen Genesis und Hybris. *Zeitschrift für Evangelische Ethik, 55*, 179-191.

Ried, J., Braun, M., & Dabrock, P. (2011). Unbehagen und kulturelles Gedächtnis. Beobachtungen zur gesellschaftlichen Deutungsunsicherheit gegenüber Synthetischer Biologie. In: P. Dabrock, M. Bölker, M. Braun, J. Ried (eds.), *Was ist Leben – im Zeitalter seiner technischen Machbarkeit?* (pp.345-369). Freiburg: Alber.

Schelling, F.W.J. (1798). *Von der Weltseele. Eine Hypothese der höheren Physik zur Erklärung des allgemeinen Organismus.* Hamburg: Friedrich Perthes.

Schummer, J (2011). *Das Gotteshandwerk. Die künstliche Herstellung von Leben im Labor.* Berlin: Suhrkamp.

Tietzel, M. (1984). L'homme machine. Künstliche Menschen in Philosophie, Mechanik und Literatur, betrachtet aus der Sicht der Wissenschaftstheorie. *Zeitschrift für allgemeine Wissenschaftstheorie, 15*(1), 34-71.

Tucker, J.B., & Zilinskas, R.A. (2006). The promise and perils of synthetic biology. *The New Atlantis, 12*(1), 25-45.

Production biology
Elements and limits of an action paradigm in synthetic biology

Tobias Eichinger

In descriptions of what synthetic biology is all about and of what characterizes this new field of biotechnology, several notions of activities are found. According to some proponents of synthetic biology, the new bioengineering discipline pursues the goal of "constructing new living things" and, in order to achieve that, "the development of tools" ("What's in a name?" 2009), as Drew Endy puts it. Similarly, for Martin Fussenegger, synthetic biology aims "to create and engineer functional biological designer devices and systems with novel and useful functions" ("What's in a name?" 2009), whereas Steven Benner expects from "building artificial genetic, regulatory and metabolic systems" the gaining of new biological knowledge (Benner 2003). Another example is a programmatic paper of Heinemann and Panke captioned "Synthetic biology—putting engineering into biology", in which the authors define synthetic biology as "the engineering-driven building of increasingly complex biological entities for novel applications" (Heinemann and Panke 2006). Tom Knight, a leading figure in the emergence of the field, enumerates a quite remarkable list of terms for identifying synthetic biology as "the intentional design, modeling, construction, debugging, and testing of artificial living systems" (Knight 2005), and Forster and Church use designations such as "genetic engineering of organisms", "constructions with new functions", "synthesis of drugs", "generate", "to create bacteria", "rebooting the bacterium", "stitching together DNA constructs", or "production of biomaterials", all of which describe activities and operations to finally "assemble a form of life" (Forster and Church 2007). As the debate about synthetic biology has been dominated in recent years by questions concerning the definition, demarcation, and identification of the field as a distinct area of science and technology that is in a crucial way a novel enterprise, a vast number of attributions and self-descriptions exist in the literature. However, the wide range of words and terms used to describe the activities and ways of working specific to synthetic biology could be divided into a few main categories of activities. These categories embody fundamental conceptual approaches and should therefore be called action paradigms.

1 The paradigm of production and engineering

Certainly, the main category dominating the self-understanding of the field of synthetic biology and its reception contains the idea of a producing and engineering biology. Among various papers, a publication by a small number of fairly prominent scientists clustered in the *Bio Fab Group* illustrates that special, technically governed concept and the engineering-driven paradigm of production. The authors, who include George Church, Jim Collins, Drew Endy, and Jay Keasling, use terms such as "to generate", "to reengineer", "to build", "to construct", "to produce", and "manufacturing" for the activities of synthetic biology (Baker et al. 2006). In this way, the new section of biology that calls itself synthetic is linked to the conceptual field of technology and technical action, or more precisely, to that of technical production and engineering. Because synthetic biology works with living entities, the anthropological and normative implications that this paradigm of human action entails are far from irrelevant.

Since ancient philosophy, modes of human activity have been the subject of theoretical investigation. Aristotle's differentiation between *praxis* and *poiesis* is an early and still influential landmark on a basic level of anthropological analysis of action theory. On the one hand, Aristotle conceptualizes forms of human activity that have their end in themselves and therefore are exercised for their own sake — these kinds of activity are cases of the Aristotelean action paradigm of *praxis*: "[A]ction and making are different kinds of things [...]. For while making has an end other than itself, action cannot; for good action itself is its end" (Aristotle 2009, 1140b/VI,5). On the other hand, Aristotle refers to the kind of activity whose purpose lies outside of itself as *poiesis*. Manufacturing and producing are paradigmatic examples of this type of activity. Thus, producing (or making) is defined by a purpose that lies outside of the action or activity but at the same time takes concrete form in the manufactured product. Producing is therefore characterized in its primary orientation or adaptation to a result whose achievement is intended. According to Aristotle, therefore, producing always means a purposeful activity that is directed at something outside of itself.

2   Purposes and interim purposes

To act in a productive way means to pursue a purpose, thereby not only implying a purpose-establishing subject, but also involving a certain degree of purposive rationality and an instrumental choice of means. At this point, it would not be trivial to mention that an end requires appropriate means for it to be achieved. Every producer is therefore dependent on having appropriate and effective means to achieve his end. Owing to the necessary production of the appropriate means, there will be inevitable 'interim purposes', or 'in-between purposes', to be interconnected before the purpose itself can be pursued and achieved. Accordingly, a product is produced along a chain of several 'interim purposes' which serve the main and final purpose of the production process for which all the *poiesis* is done. Within the course of the production process — which takes place only in a mediate way — the 'interim purposes' themselves have the function of subordinate purposes.

A remarkable characteristic of the particular purpose-means or end-means relationship in processes of production is the fact that not only can means become ends on their own, but moreover, that ends may also be pursued as means to another end. That is to say, the finished products obtained at the end of a production process may, in turn, be intended to serve another purpose which is only achievable in a mediate way. Thus, the technical product contains "action potentialities": "Production creates mediating action potentialities which are to be updated in the application. Technical action is, in its characteristic entanglement of production and application, intermediate action [...]. Producing action is never other than a means to an end, which fulfills itself only in the course of applying action" (Ropohl 1996, p.91)[1].

This notion of intermediacy — that achieved ends can themselves serve as means for further ends — is incarnated in a paradigmatic way in the production of tools, machines, and the like. Tools and machines are objects that are basically only pursued in order to achieve another, subsequent end in a better way, or to achieve it in the first place. In this respect, they are 'artifacts for the production of artifacts'. In an anthropological turn, the capacity for instrumental invention and systematic production of means — that is, the production of tools — could be qualified as a central feature of human nature. Benjamin Franklin expressed that concisely in his well-known notion of "man as a tool-making animal".

---

1   Own translation. German original: "So schafft die Herstellung mediale Handlungspotentiale, die in der Verwendung aktualisiert werden. Technisches Handeln ist, in seiner charakteristischen Verschränkung von Herstellung und Verwendung, intermediäres Handeln [...]. Herstellungshandeln ist immer nur Mittel zum Zweck, der sich erst im Verwendungshandeln erfüllt".

With respect to the purpose orientation pervading the complete course of a production process, Hannah Arendt follows the Aristotelean conception. She emphasizes the difference between acting (*praxis*) and producing (*poiesis*) (and also between labor and production) which lies in the fact that the latter comes to a closure with a defined end product: "In the process of making […] the end is beyond doubt: it has come when an entirely new thing with enough durability to remain in the world as an independent entity has been added to the human artifice" (Arendt 1998, p.143). The product is added with its materialness or objectness to the inventory of the world, where the final result provides proof of evidence for the occurrence of the production process in a concrete and tangible way. According to Arendt, the attribute of a clearly determinable finish and closure defines production and producing as a unique capability of man: "To have a definite beginning and a definite, predictable end is the mark of fabrication, which through this characteristic alone distinguishes itself from all other human activities" (Arendt 1998, pp.143f.).

3   Durability and dependency

Taking the finality of producing (or in Arendt's words "fabricating") and the objectness of the products together, it can be seen that, once produced, the outcomes of the process of production could persist on their own without being applied to their original purpose. Fabricated products have become part of the non-human world as objectified purposes. And just like the rest of the non-human world, they are accessible and free for any usage, whether for reasonable and meaningful applications, for purely senseless uses or for counterproductive abuse. The product's discreteness and self-reliance in terms of its complete and arbitrary availability originate in principle from the relation between the product and its producer. Just as each manufactured product is determined by the purposefulness of its existence by its being produced on condition of its intended ability, so is the purpose of the process of production or fabrication determined for its part by the producing or manufacturing subject. As long as the process of production is still ongoing the product, i.e., the thing to be produced, is totally dependent on the producer. During the course of fabrication the final product does not yet exist, at least not in material form but only in an abstract or virtual form of the initializing idea. What exist, apart from the idea, are the components or basic materials from which the later product is to be assembled and fabricated. The product takes its final shape only by and by in the course of its process of origination, which is entirely subordinate to the will and power of the producer.

At the moment in which the intended final stage of the production process finishes, the product's total dependency on its creator is abruptly severed. The produced object is now released into its self-reliance and the durability of an objectified existence. Indeed, having been conceived of, designed, compiled, and completed in a very purposeful manner, the produced object unfolds on its finalization attributes of an independent entity by which it is inevitably contrasted with the producing subject.

But there lies a certain tension or ambivalence. On completion of the production process the product is a finished product in its own right and no longer dependent on its manufacturing process because it has been produced to endure. Certainly, that does not imply that it will or should endure. Its enduring nature, or persistence, is determined by another characteristic of the concept of production, which is that the basic attributes of production and the produced object — concreteness and durability — and the central element of purposefulness give rise to: predictability and controllability. As the production process and product are predictable and calculable, they are directly opposed to the principle of natural evolution based on chance and unpredictability. The producer is the sole originator of both the production process and the products, which would never have come into existence without him. As a consequence of that dependency, the producer has total control over the fate of his products — which could also imply a kind of reversion of the production process, i.e., the possibility to destroy the constructed product, as Arendt writes: "This great reliability of work is reflected in that the fabrication process, unlike action, is not irreversible: every thing produced by human hands can be destroyed by them, and no use object is so urgently needed in the life process that its maker cannot survive and afford its destruction" (Arendt 1998, p.144).Thus, products are characterized by being durable, or more precisely by having the potential to endure — albeit not the certainty of doing so.

The product idea presents a slightly different case. Like the product itself does — at least potentially — the idea behind its production, which includes a purpose definition, outlasts the accomplishment of a production process. What really outlives the production procedure is the underlying model and draft. The endurance of the production draft gives rise to another central attribute of the production or fabrication paradigm. Arendt writes: "the image or model whose shape guides the fabrication process not only precedes it, but does not disappear with the finished product, which it survives intact, present, as it were, to lend itself to an infinite continuation of fabrication"(Arendt 1998, p.141). The vision or idea behind the production process is in its abstract anticipation completely independent of any realization. Therefore an infinite number of entirely unconnected production procedures could refer to a single original idea, to the very same source draft, and could of course be repeated again and again. Thus, the fundamental potential for duplicating and mass producing in any manufacturing procedure — of the process itself as well as of the product — is already contained in the concept and structure of production. What has been produced could in principle also be reproduced in any quantity (disregarding potential limitations of existing resources and other factors external to production).

4 Anthropological aspects

In the same way that the everlasting idea and model of a production offer the possibility of constant realization and thereby persist, so do the results of production, which detach from their producer as soon as they are produced, have the tendency to be permanent results. Notwithstanding the ever-existing possibility of destruction, we have here a very important feature from an anthropological perspective. Accordingly, Arendt emphasizes the attribute of the produced object that is capable of outlasting its producer using the notion of durable: "It is this durability which gives the things of the world their relative independence from men who produced and use them, their 'objectivity' which makes them withstand, 'stand against' and endure, at least for a time, the voracious needs and wants of their living makers and users" (Arendt 1998, p.137). Processes of production never remain without results or marks in the world. That effect occurs in the objectified form and tangible concreteness of the produced products. According to Arendt, the effect of this materializing and objectifying process is not limited to the manufactured objects themselves, "because they give the human artifice the stability and solidity without which it could not be relied upon to house the unstable and mortal creature which is man" (Arendt 1998, p.136).

The notion of "to house"—which is the translation of "Heimat" or "beheimaten" in the German version, meaning "home" or "to be native in the world"—contains a fundamental anthropological meaning in that man creates a "house" for himself by objectifying things in the course of his own activity of production. The unsettled and frail human being, whose life in its structure and progress is in a permanently vulnerable state and constantly inconstant, has an existential need for a certain degree of stability and durability. Here we can identify a classical concept of philosophical anthropology which conceives of man as a free being treated in a novercal or stepmotherly way by its mother nature (the topos of the *natura noverca* which goes back to the ancient Greek and roman philosophers, prominently readopted in the age of Enlightenment by Johann Gottfried Herder and taken as a basis of the anthropological conception of man as a "deficient being" by the twentieth-century

German philosopher and sociologist Arnold Gehlen (1988)). That image of man as a permanently threatened being who requires, in order to be viable and survive, the ability to complete his own unnaturalness by self-made artificiality — and therefore by acting as *homo faber* — is a notion Arendt uses explicitly: "Work is the activity which corresponds to the unnaturalness of human existence [...]. Work provides an 'artificial' world of things, distinctly different from all natural surroundings. Within its borders each individual life is housed, while this world itself is meant to outlast and transcend them all. The human condition of work is worldliness" (Arendt 1998, p.7).

5 Production biology

These more general features of the pattern of production indicate that the production paradigm involves a number of aspects that are predominant for synthetic biology in conceptual and practical respects. One aspect is the relation of synthetically produced objects to nature. The purpose of a produced product and its production process is (usually) a human purpose, for the product would typically not have come into existence without human action: "What is constitutive for technical action are real objects not realised without human intervention in the encountered natural world" (Ropohl 1996, p.85). This statement fits in with a very common definition of synthetic biology that stresses "the design and construction of new biological systems *not found in nature*" (Markus Schmidt 2009, p.1). The specific technical action in synthetic biology leads to results that consequently extend the range of nature and natural objects. According to this idea, a certain removal from nature if not resistance against it is also a typical feature of every produced thing. Thus, the notion of artificiality and of artifacts involves also the concept of remoteness from nature. Of course, since the objects produced in synthetic biology are living entities and systems built out of natural parts and modules, a special kind of artificiality is present here. It is not directly opposed to the sphere of nature as is the case with classical engines, machines, electronic devices or other non-living technical instruments and gadgets. In synthetic biology, removal from nature concerns the processes of origination rather than the consistency of the results. This circumstance accords with the distinction between genetic and qualitative naturalness (Birnbacher 2014, p.7). Whereas genetic naturalness concerns the way in which something has originated, qualitative naturalness (and also artificiality) indicates a certain constitution and appearance. In the latter sense, the objects produced and created by synthetic biology are in the majority of cases as natural as living objects and systems found in nature. What is different and constitutes as being new, innovative, exciting or groundbreaking, particularly in synthetic biology, is the process of their development or fabrication. They are simply products of engineered biology. And because they exhibit a categorically different genesis than naturally grown objects it is an open question whether they contain, or whether they should be assigned the same value only because of their qualitative naturalness. This question leads to the foundation of ethics and is impossible to answer without taking a stance on the source of normativity. Ultimately, the moral status of the objects of synthetic biology depends on the fundamental ethical position one takes, whether from an anthropocentric, pathocentric, biocentric or other angle.

Another aspect of this special amalgamation of qualitative naturalness and genetic artificiality is more relevant for determining the peculiar characteristic of synthetic biology as an engineering discipline. The fact that the material engineered by synthetic biology is both biological and natural puts another complexion on the defining paradigm of production and its implications. It makes the engineered substance a very special one: a substance that not only complies with the interventions and manipulations of its manufacturer, but also has a certain degree of independence. The engineered objects of synthetic biology exhibit such independence because of their naturalness, that is, their liveness. By their being alive, entities, objects or systems of any size or on any level have a stake in the dimension of a kind of natural originality and self-will. They are not inanimate, but things that could react in an unexpected way and are capable of Darwinian evolution. This circumstance demands a particular approach and way of dealing with synthetic biology's paradigm of production. The engineering, synthesizing and fabricating of artificial life forms can only be successful if it counts on the special intrinsic dynamics of living matter. It is precisely that characteristic which makes synthetic biology special and novel, as Matthias Heinemann and Sven Panke stress. They perceive "a fundamental difference between engineering biology and engineering in other natural sciences such as chemistry or physics." The fundamental difference lies, according to these synthetic biologists, in the fact that "biological systems have the capacity to replicate and to evolve" (Heinemann and Panke 2006) – in other words, the capacity to live. Hence, synthetic biology as an *engineering natural science* has to face difficulties of unexpected and — to a certain degree — unpredictable outcomes. Thus, it is bound to that peculiar rest of life that defies control.

6 Tinkering production

From a methodological perspective, such difficulties were bound to have consequences for practical approaches. And so it is no coincidence that some scientists in the field of synthetic biology point to the need for revising the way their activities are described and conceptualized, and therefore for a change in their self-perception. What is meant is the need to broaden the description of their range of activity by the term *tinkering*. Petra Schwille has published a programmatic paper entitled "Bottom-Up Synthetic Biology: Engineering in a Tinkerer's World" (Schwille 2011). Steven Benner and colleagues have done similar using the term "tinkering biology" (Benner et al. 2011). The suggestion is to supplement the methodological procedure with the principle of trial-and-error, owing to the specific qualitative naturalness of the objects concerned. Tinkering refers to a way of responding to and cooperating with living matter and its contingency, rather than to determining and commanding it. Synthetic biologists engaged in tinkering also acknowledge certain limits of prediction that in principle emerge when working with living objects. That does not mean that the production paradigm is misleading in characterizing synthetic biology, or even that it could serve as an appropriate leverage point for ethical criticism of the whole field. In its classical form, the action-theoretical and anthropological analysis of producing and fabricating, as Hannah Arendt presented it, does not neglect the activities in synthetic biology, but it seems to be too short-sighted. The strong notion of accurate predictability and absolute controllability (up to destructibility in normative respects) has to be restricted and complemented by the openness and incalculability implicated by the action mode of tinkering. At the same time, this cannot mean blind and wild synthesizing and re-combining by accident over and over again, as Benner et al. emphasize: "Nothing of value comes unless the tinkering is followed by studies of what happened" (Benner et al. 2011, p.374). Of course, in the context of a highly engineering-focused high-tech science every little step of failure has to be analyzed and learnt from to avoid repeating it in the future. But the paradigm of *tinkering production* assumes these limitations of living and natural matter from the very beginning. That this attitude is ultimately unavoidable might be illustrated by an aspect that is a central element of almost all discussions, papers, hearings, statements or reports dealing with the social and ethical issues in synthetic biology: the subject of biosafety. No serious scientist in the field or observing researcher would probably deny the importance of biosafety issues in synthetic biology (cf. exemplarily Schmidt et al. 2009). This shows that there is a strong awareness of the limits of control over living entities. From an ethical standpoint, it does not seem exaggerated to conclude that doing synthetic biology in a tinkering manner reinforces rather than weakens the respect for and appreciation of living matter. But most notably, it demonstrates the appropriateness and importance of the tinkering concept for an adequate characterization of the specific engineering and production paradigm in synthetic biology. This paradigm sets production biology apart from traditional manufacturing and producing technologies by synthesizing the two conceptual spheres of nature, evolution, and life on the one hand and engineering, producing, and fabricating on the other.

Literature

Arendt, H. (1998). *The Human Condition*. Chicago: University of Chicago Press.
Aristotle (2009). *Nicomachean Ethics* (trans. by W.D. Ross, ed. by L. Brown). Oxford: Oxford University Press.
Baker, D., Church, G., Collins, J., Endy, D., Jacobson, J., Keasling, J., . . . Weiss, R. (2006). Engineering Life: Building a FAB for biology. *Scientific American, 294*(6), 44-51.
Benner, S.A. (2003). Act natural. *Nature, 421*(6919), 118. doi: 10.1038/421118a.
Benner, S.A., Yang, Z., & Chen, F. (2011). Synthetic biology, tinkering biology, and artificial biology. What are we learning? *Comptes Rendus Chimie, 14*, 372-387.
Birnbacher, D. (2014). *Naturalness. Is the "Natural" Preferable to the "Artificial"?* (trans. by D. Carus). Blue Ridge Summit: University Press of America.
Forster, A.C., & Church, G.M. (2007). Synthetic biology projects in vitro. *Genome Research, 17*(1), 1-6. doi: 10.1101/gr.5776007.
Gehlen, A. (1988). *Man. His Nature and Place in the World* (trans. by C. McMillan, & K. Pillemer). New York: Columbia University Press.
Heinemann, M., & Panke, S. (2006). Synthetic biology – putting engineering into biology. *Bioinformatics, 22*(22), 2790-2799. doi: 10.1093/bioinformatics/btl469.
Knight, T.F. (2005). Engineering novel life. *Molecular Systems Biology, 1*(1). doi: 10.1038/msb4100028.
Ropohl, G. (1996). Technisches Handeln. In: G. Ropohl (ed.), *Ethik und Technikbewertung* (pp.83-108). Frankfurt/Main: Suhrkamp.
Schmidt, M. (2009). Introduction. In: M. Schmidt, A. Kelle, A. Ganguli, & H. d. Vriend (eds.), *Synthetic biology. The technoscience and its societal consequences* (pp.1-4). Dordrecht: Springer Academic Publishing.
Schmidt, M., Ganguli-Mitra, A., Torgersen, H., Kelle, A., Deplazes, A., & Biller-Andorno, N. (2009). A priority paper for the societal and ethical aspects of synthetic biology. *Systems and Synthetic Biology, 3*(1-4), 3-7. doi: 10.1007/s11693-009-9034-7.
Schwille, P. (2011). Bottom-Up Synthetic Biology: Engineering in a Tinkerer's World. *Science, 333*, 1252-1254.
What's in a name? (2009). *Nature Biotechnology, 27*(12), 1071-1073.

Creativity and technology
Humans as co-creators

Harald Matern

Foreword

In this chapter I propose to offer a genuinely theological perspective on synthetic biology. It is a perspective with ethical implications. However, I will not be concerned primarily with issues of application, but will discuss a fundamental discourse issue, from which suggestions for specific actions may nevertheless be taken. Hence, the intended perspective is one of dogmatic and religious-philosophical relevance and concerns especially theological anthropology and the doctrine of creation.

Following an initial introduction to the topic, including an outline of the historical theological background, I will discuss key concepts – or metaphors – from the field of synthetic biology that are relevant to the theological discourse in anthropology and the doctrine of creation: 'life' and 'new'. The question at issue is whether synthetic biology puts the case that human beings must be seen as creators of life. That would, at first glance, pose a theological problem. Moving on, I will argue that, contrary to what some public debates appear to indicate, Protestant theology has a long affirmative tradition that essentially aims to mediate the relation between divine and human activity or working with regard to what 'creation' 'is'. If such a 'co-working' can be considered a fundamental anthropological aspect of the doctrine of creation, it will be possible to arrive at a tighter definition of what it means to speak of humans as 'co-creators' with the aid of the concepts of 'creativity' and 'technology'. After this narrowed-down analysis, I will conclude by addressing the ethical implications of such an anthropology and renew the question of whether, and if so how, the latter is fundamentally challenged by synthetic biology.

1 Introduction: Creating life and playing God?

A good part of the hopes and fears associated with synthetic biology in the public's perception seem to revolve in particular around two terms: 'life' and 'creation'. Is life created in synthetic biology (Schrauwers and Poolman 2013)? Is synthetic biology then a "divine craft" (Schummer 2011)? And if so, what does that mean, and what implications would it have for the image of humanity and of the world (Boldt et al. 2009; 2012a; Dabrock et al. 2011)? Would it not mean that humans are "playing God" (Peters 1996; Dabrock 2009; Dabrock and Ried 2010; 2011)? And precisely what kind of activity is being described (Boldt et al. 2012b)?

These questions are very relevant because they have a bearing on fundamental aspects of human self-perception in contemporary Western societies. And they cause a bigger response in the media than others because they cannot be completely rationalized. This circumstance is due, on the one hand, to the concepts used. 'Life' and 'creation', or 'creative working', should perhaps be called metaphors rather than concepts, meaning that they cannot be fully defined, and the semantic surplus they contain carries a certain communicative pragmatism: the use of metaphors is not only an expression of a cognitive "quandary", but often intentional and appellative (Lakoff and Johnson 1980; Blumenberg 1997; Haverkamp and Mende 2009). On the other hand, this linguistic peculiarity also refers to the phenomenality of 'life', to the fact that it describes something that needs and in a certain way eludes linguistic access at the same time (Thompson 2008; Fellmann 2010).

In theology, life is associated with God in a primal sense. God is the creator and sustainer of life; God creates life 'out of nothing'. Likening oneself to God is considered to be humanity's original and hereditary sin. The process that comes closest to the divine creation of life – human reproduction – was linked with sin at a very early time in church history. Not without reason does the suspicion someone is planning to 'create life' and thereby 'play God' also cause a theological stir, although theological debates on this topic did not first arise in the context of the discussions on synthetic biology. Rather, it was genetic engineering that raised the question of humanity's undue claim of creativity in the 1980s, prompting at first academic-theological statements of opinion (see, for example, Wehowsky 1985; Hübner 1986; Dohmen 1988; Bender and Gerber 1990), and then explicit church statements (cf. EKD 1991). The debates, for their part, had arisen in connection with theology's response to the general ecological sense of crisis that had infiltrated into theology in the 1970s (cf. only Améry 1972) and provoked reactions following the publication of the Club of Rome's report "The Limits to Growth" (Meadows 1972) and through the connection between Christianity and ecological destruction (White 1967; 1973).

For the theological discourse it is not of minor importance that religious semantics also played a role in public debates. As early as in 1991 Christian Link maintained that the "problem with creation" had in no small measure and "in a stormy development migrated from the inner room of theology into the public discussion on science and worldviews" (Link 1991).

The precise nature of the "problem with creation" becomes clear on closer examination of the contexts in which the described 'migration' took place. 'Creation' is becoming a programmatic word of a moral ecological debate on the unity and intrinsic value of the world under the aspect of the life existing in it (cf. in a critical vein Graf 1990). The migration of precisely religious semantics indicates a particular quality of this imagery, which may serve to capture what requires justification but is difficult to justify so that the significance of that being described is also articulated for the observer, and speaking is not a purely subjective expressive action.

The observation of a 'migration' of religious semantics into the public discourse remains valid – particularly with regard to synthetic biology. It is still true that "no other lesson of Christian dogma [...] has conquered such an important place for itself in everyday language than the ecologically-emancipatorily reshaped doctrine of creation" (Anselm 2012, p.233).

That said, the current situation is witnessing a specific aggravation and intensification of the debates. The discussions do not focus on a religious underpinning of a general ecological commitment, or on the 'preservation of creation' in general. It is rather that synthetic biology raises very specific issues which go hand in hand with the use of a specific scientific terminology. It would be a fatal mistake, also from a theological perspective, to forget the difference between religious and non-religious semantics, and to ignore that 'creation' is not a concept in the proper sense, but figurative speech.

I have shown above that the 'migration' of religious semantics into the public discourse occurred in a context that put Christianity and its theology in an apologetic position. In theology, this circumstance led, at least for a short period, to a chiefly critical view of humanity's role in the stability of the ecosystems. Technology was also seen in a largely critical light in the early technological debates on genetic engineering. Likewise, the use of religious semantics in the public sphere bears this stamp. The situation appears to be no different in the case of synthetic biology: Expressions such as 'playing God' or 'creating life' initially stir up vague fears or possibly moral indignation in the face of such 'pretensions'. Once the ecological issues had been inscribed into creation theory thinking, the critical view of humanity and the ecologically charged semantics of creation moved out of the theological debates and returned to the public domain. Evidently, we are faced with a distinct discursive interdependency between religious and non-religious communication contexts. A further task of theology is therefore to reveal and critically reflect on this relationship if the intention is to gage the implications and potentials of theology's handed-down metaphors in this debate.

That is particularly important with regard to debates on synthetic biology. In light of the (theological and non-theological) fears and condemnations of humanity's self-perception surrounding the discourse on synthetic biology, it is easy to forget that Christianity and Christian theology have a long tradition of affirming humanity's creativity. In modern times this tradition has been engaged in an ongoing exchange with secular traditions. While the modern tradition of the creative human being is usually associated with Pico della Mirandola (cf. Tauber 2001), from the theological perspective it was Martin Luther who basically presupposed a creative co-operation of God and humanity (cf. Ebeling 1989, in particular section 62 on the Cooperatores Dei; Joest 1967; Härle 2011). At the same time, Luther restricted such co-operating conceptually by stating that human beings are God's "cooperatores, non concreatores" (Luther 1912, p.857, 35). In doing so, a problem that has occupied the Christian doctrine of creation until the present day was born. The principal concern is not with drumming in the idea that God is the creator and man is the creature; or that God is God but man is a sinner, as Wolfgang Huber argues (1990, pp.189f., passim) – and that any pretension of creativity must be immediately condemned.

Rather, the crucial problem lies in the description of the co-operating of God and humans with respect to the world as 'creation', and in how the relation of relative unity of God and humans is portrayed without ignoring their difference. Naturally, the discourse underwent change as a result of humanity's expanded knowledge of the world through scientific and technological progress, but in particular through its expanded technical possibilities (cf. Matern 2015 for a detailed overview).

The biotechnologies pose a particularly huge challenge to theological reflection because they really do unlock new potentials which are not easily conceived by global metaphors such as 'preserving creation'. And it is not without reason that the debate surrounding synthetic biology also has a particularly explosive character in terms of theology. Although synthetic biology can be regarded as a continuation of genetic engineering in many respects, its focus is, more clearly than before, on the fabrication of life forms that are 'new' in a qualitative sense. The qualitative leap forward is made possible by reorganizing the biotechnological possibilities by means of engineering science.

2 Historically new – new life?

As a consequence, the important criterion for theology is not only the concept of 'life', but in particular the concept of 'new' (cf. Dabrock et al. 2011). The latter is intentionally related to the concept of 'nothingness', because it represents the temporal counterpart of that ontological category. Something new is something that has historically not yet existed. Something radically new cannot be explained causally, nor can it be represented temporally as a consequence of a development already begun. The genuinely new thing itself is a beginning and origin of something else.

Control over the 'new' therefore has an almost apocalyptic quality – in the truest sense of the word. The new object can be regarded as the disclosure of what is not yet known, as the unveiling and revelation of a transcendental reality – or as a chance event. Each case describes a conceptually uncontrollable process, which is now acknowledged as being practically realizable. In theological tradition, the ability to realize something that is radically new is ascribed to God alone, not to human beings. A radically new thing is something out of nothing. It goes beyond human capacity for explanation or action. Producing a genuinely new thing is a *creatio ex nihilo*, a creation from nothing. In the Christian dogma, novelty is a category that could be ascribed to the doctrine of creation and to eschatology in equal measure (Sass 2013).

In recent biology and biophilosophy, however, one finds a thought model that does not attribute a new entity to an action. According to this model, the coming into existence of a new entity is not causally linked to intentional intelligence, but viewed as an epiphenomenon of other processes. The process by which something new arises 'from nothing' is known as 'emergence'. From a theological standpoint, God's working can still be assumed in such processes so that the epiphenomenally emergent entity can ultimately be seen as the outcome of a will of some kind and thus of a purposeful development (Thomas 2013). Of course, such a perspective gives rise to new problems. One particularly relevant question concerns the existence of a set of criteria that would allow highly valued new entities to be distinguished from the not so desirable ones. These problem areas also pertain to an ethics that is mainly guided by the concept of life (cf. Schweitzer 1966, Gansterer 1997). Here again, the question of how the partly contra-intuitive linking of the 'good' with 'life' is to be not only comprehended but also operationalized begs to be answered.

Accordingly, a more precise definition of the concept of 'new entity' or 'new life' in the field of synthetic biology is of interest. Such definition will also determine how one can theologically grasp the role of the bioengineer who would also be involved in producing 'new life' in this context.

How really new is the 'new entity' in synthetic biology – what is its specific nature? Is 'life' that emerges in connection with a synthetic genome new simply because the genome was completely synthesized – although as a replica of an existing genome? Should the attribute of new only apply from the point at which the genome was modified in order to bring forth hitherto unknown functions in life forms? Or does the new only begin where possible life forms emerge on a previously non-existent basis, where viable life not based on DNA exists? The experiments directed at XNA life forms – as well as xenobiology – are open to the possibility of extending the already fragile attempts to define 'life' to include what has so far been omitted. As a side note, it is worth noting that this process would entail significant logical problems (how can a concept that is still to be precisely defined be extended by an empirical approach?). Is the concept of 'new' therefore first and foremost an epistemological problem, and not an empirical or ontological one (echoing the view of Sass 2013)?

It seems clear that the 'new' in synthetic biology is to be found on at least two levels. First, in the sphere of actions: the combination of molecular-biological and chemical procedures with engineering is what makes up the specific character of synthetic biology in this respect. Elsewhere we have endeavored to show that the actions of a synthetic biologist come much closer to product design than to an artistic, "creative" act of creating (Boldt et al. 2012b). At the same time, a categorical difference compared to current biotechnologies, which would allow one to speak of a qualitatively new technology, cannot be identified here.

On a second level, one would need to query whether the product as such can be deemed 'new'. Evidently, when considering 'new life' a strict distinction must be made between life forms having hitherto non-existent or unknown functions and life that is truly 'new' because it is 'different'. If it is a case of life whose fundamental characteristics make it clearly different from known life, calling it new is certainly appropriate. However, neither life forms having new specific functions nor even XNA-based life are so radically different from known life that they could be classed as being ontologically new in the sense of emerging 'out of nothing'.

On closer examination, something else is disconcerting for the observer. And it is the observer's discomfiture that is crucial both theologically and morally. First, the phenomenality of life is basically incompatible with reductionist-technical and mathematical perspectives. The reason is that 'life' has to be attributed to 'living matter' at the very point at which its self-referentiality is manifest. However, anything capable of referring to itself is resistant to an analytical approach. Its resistance is due to the fact that the observer sees precisely therein a similarity to him or herself (cf. Gerhardt 2002, p.317). At this point, it is not about the sight of another, which basically implies the supposition of reciprocity. In a much more basal sense, life creates the impression of such autonomy even at the cellular level. The notion that 'new life' could be produced in a purely technical way from inanimate matter is particularly disconcerting because in this image 'life' is deprived of its resistance. This statement applies in particular to approaches that link a constructivist perspective on 'life' to a constructive-linear research approach (cf. on this subject Witt 2012).

With that, attention is once again turned to acting human beings. All at once, the scientist or bioengineer is placed in the role of 'creator'. Anyone who 'takes hold' over life technically deprives it of its ontological independence at the same time. The ensuing discomfiture for a systemic-organologic understanding of life is expressed in metaphors such as 'playing God'. Not without reason does the notion that humans would be creating 'life', or more precisely 'new life', cause certain unease (Ried et al. 2011). The unease and discomfiture is reflected, for one, in the medial use of religious semantics to describe the situation. The metaphorical or religious semantization of those processes that affect the foundations of the lifeworld-oriented self-understanding of Western modern society contain expressions of fears and intellectual difficulties, confusions regarding certain concepts effective in everyday life. Are the processes in the context of synthetic biology perhaps indeed a "divine craft" (Schummer 2011)? Such a statement probably does not correspond to the self-perception of the majority of scientists involved. And yet, it points to an important factor. The question of 'novelty', and in particular of 'new life', exhibits a scientific-ethical dimension besides an ontological-metaphysical and biological one.

This dimension of synthetic biology and its public perception is particularly relevant for theology. The ecological-emancipatory creation theology of the 1970s and 1980s, with its critical perspective on humanity, specified very one-sided uses for the metaphor of 'creation'. It is frequently forgotten, even in theology, that a large part of contemporary creation theology contradicts such a specification. Instead, the theological doctrine of providence has engendered a thought tradition whose object is to mediate the relationship of the different agents in the emergence of something historically 'novel'. How can the relationship of 'God', 'man', and 'nature' in its practical dimension be represented logically and linguistically? How can the continuation of history and the emergence of novelty be interpreted in terms of the collaborative work of these different agents?

This implies that contemporary theology of creation displays a tradition of reflection which has shaped a clear boundary consciousness, a special attention to both the ontological-metaphysical and the practicable fragility of dealing with the emergence of 'novelty'. Falling back on this aspect of traditional theology of creation is very promising from an ethical standpoint because it constantly encourages the avoidance of one-sidedness. Neither excluding the religious valency of human self-activity nor exaggerating human activity, or even its deification, can be meaningful in terms of theology of creation. Mediating between these two poles is a complicated matter, especially where the question of ends and outcomes is concerned.

In the following paragraphs, I will briefly elaborate on the above.

3 Collaborative working: creation as teamwork?

The theology of creation is traditionally treated in two major parts. While the theological concept of *creatio originalis* is concerned with the creation of the world out of nothing before the beginning of time and with the creation of animals and humans, the heading *creatio continua* refers to God's ongoing interaction with the world and humanity throughout history. This concept correlates with the biblical narratives in Gen 1:1-2:4a and Gen 2:4b-25, even though they differ substantially. In the first account the world is created by the Word, whereas the second text describes the creation of humans and animals from the earth by a creative act. Evidently, it is assumed that God acts in different ways, particularly with regard to the creation of living matter. In both texts human beings are given the role of tending and preserving what God has created (Zenger et al. 1995; Link 1997; Schmid 2012).

Disregarding all the differences, human beings have thus been given a very important role. As God's representative and by God's order, humans are to keep and cultivate the world and all living things existing in it (animals and plants). In the traditional dogmatic account of creation, the relation between divine and human activity, or operating was relevant particularly with regard to this latter issue. Evidently, a co-operation of God and humans has to be assumed with respect to the history of the world. This aspect has traditionally been described by the terms *concursus* and/or *cooperatio*. One difficulty that arose concerned the distinction between humans' act of cultivating and activities going beyond that framework. According to the Bible, human beings were punished for overstepping such bounds by being expelled from the Garden of Eden (cf. Gen 3). Hence, the doctrine of creation and the doctrine of sin belong together, although the borderline between cultivating and sinful human activity was blurred from the beginning.

Such distinction did not pose a particular problem until the modern era. Before the Enlightenment in Europe, theologians could generally assume God's divine intervention in the course of historical events, however this notion became increasingly doubtful with the growing criticism of classical metaphysics and the biblical stories of miracles. The adoption of this criticism in theology was first expressed in detail in the work of the theologian and philosopher Friedrich Schleiermacher in the first half of the nineteenth century (Danz 2007, pp.95-98,113-125). Two aspects are particularly significant in this respect. First, God ceases to be conceptualized as an acting subject in the creatio continua. God's agency is henceforth rather described as an affecting than an acting. Second, a third independent agent – namely nature, which had previously been mostly treated as being pure matter – comes on the scene (cf. Koch 1991; Barth 1995).

Thus, a constellation that was to remain relevant for theology of creation in the modern age was established. It is articulated in the semantic shift from 'acting' to 'taking effect': an intervening, acting God was at first no longer a topic, but certainly the interdependency between humans and their historical nature, within which divine interventions could hardly still be identified as such. Human (artistic) creativity could be associated with the artful establishment of the world by God (Käfer 2004). What had been debated in philosophical discourse since the Renaissance had now made its way into the theological doctrine of creation: the creative human being. Simultaneously, however, a unique form of creativity was also ascribed to nature. The emergence of a new living entity was viewed as the outcome of natural processes, whereas culturally novel things were attributed to humans. Divine intervention was understood as a human interpretation, not as an empirically identifiable process.

The described basic constellation did not undergo further fundamental change and necessitate a revision of theological thinking until the arrival of the more recent technologies, in particular the biotechnologies after World War II. After the holocaust, God's presence in history seemed hardly conceivable. And the fact that humanity's means had increased to such an extent that basically the complete and even multiple eradication of humanity was now within the realms of possibility ('overkill') posed fundamental problems for theology of creation. Finally, the possibility of destroying humanity's bases of existence by the exploitation of resources and destabilization of the ecosystems also fueled a sense of crisis in ctheology of creation.

The major debates in theology of creation responded to these changes by making categorial adjustments. In doing so, they came back in a certain sense to Schleiermacher's approach, while neutralizing its constructivist edge. Theology of creation now addressed the interdependency between humans and nature within the world. But in the second part of the twentieth century divine works were no longer relevant exclusively as the results of human processes of interpretation. Instead, God returned into the world. The notion of God was inscribed in different ways in historical and natural processes. The assertion was certainly realistic. A new metaphorism allowing a redefinition of God's 'working' was also used. One particularly prominent metaphor was that of the "force field" (Pannenberg 1991, p.122), by which God's 'operating' in the world could be conceived of in the fashion of a physical phenomenon. "Sphere of potentialities" (ibid.) became another popular way of describing the immanence of God in the history of nature. The "co-operation" of God, humans and nature was even emphasized to such an extent that the assumption of an original creation of the world "ex nihilo" was entirely abandoned (Welker 1995).

Consequently, humanity and human responsibility take up a central position within the theology of creation. The theology of creation is itself definitively 'ethically' charged. However, contrary to most of the concepts popular in the nineteenth century, anthropological reductionisms are now to be avoided if possible. Hence, the new emphasis on divine immanence in historical nature is not intended to deprive the latter of its autonomy or to undermine the natural scientific explanation of the world. Such concepts did not of course, despite their realism, aim at returning to pre-Enlightenment times from an epistemological standpoint and to conjure up a mythical worldview. Hence, the new realism was also labeled "critical" and subjected to appropriate epistemological debates (cf. recently Losch 2011; 2014; Link 2012; Wright 2013). On the contrary, the concept of a natural world once again filled with divine presence made it possible to deal with the new ethical issues in a completely new way. By portraying nature as humanity's counterpart and as a place of divine presence, human beings could also be viewed in a new light as part of the natural world. Responsibility for humanity and for the world was thus conceived of as belonging together.

It is true that genetic engineering first triggered a fundamental debate on technology in theology. This no doubt came about because the borderline between natural and human creativity appeared to be becoming blurred. The independence of historical natural processes seemed only now to be so fundamentally in doubt that it was hardly still possible to deny that human beings possessed godly-creative abilities. A major point of concern in this regard was the question of the 'creatability' of humans by other humans (cf. by way of example Altner 2001; Huber 2002; Weth 2004; Körtner 2005). Nevertheless, the ecological issues did not go astray. It was supposed to be possible to understand technical activity, analogous to divine 'creating' in the context of the *creatio continua,* as 'co-operating'. This concern is articulated, for example, in the concept of "alliance technique" which is popular in the German debate (Moltmann 1987, p.57; see for a more general description of the concept: Nordmann 2007).

A further point is significant here. Wherever God was inscribed in the history of nature, an *entelechy* was encoded in the latter. The end was not to be given to 'the creation' first of all by humans; it was actualized in creation itself from the very beginning. All of the new theologies of creation were eschatologically charged, meaning that the traditional notion of fulfilment at the end of history was here inscribed into the conception of history itself. At the same time, however, the end was not supposed to be the result of a linear process within which humans culturally reshape the natural world. Rather, humans and nature alike were seen to be caught up in such a process of transformation. Eschatology consequently has a dual function in this case. On the one hand, it allows the description of human and non-human nature in a mutual process, thereby making a 'co-operating' of man and nature conceivable. On the other hand, the 'eschatological reservation' highlights that the realization of the end is removed from active access in a dual sense. First, in a temporal sense: the fulfillment is still outstanding. Second, in an ontological sense: the subject achieving the fulfillment is not humanity, but God. That God works in the world and not without human or non-human nature, remains notwithstanding in these conceptions.

The described transformation of the theology of creation in the modern age has been succinctly articulated by the North American Lutheran Philip Hefner. The latter's conceptualization (for an overview cf. Ried 2011) caused quite a stir in Anglo-American theology on account of his deliberately affirmative attitude to human beings' creative capacities. His concept of a theology of creation does not only reckon with the immanence of God in the natural historical process. Rather, through his affirmative reception of the theory of evolution he is able to give human beings a role that goes far beyond that of a catalyst of certain processes. Hefner assumes that the human capacity for culture and technology represents a turning point in evolutionary history. Accordingly, humanity has been tasked with driving the process of evolution forwards by means of purposeful, creative interventions. Such human destiny is at the same time directed at God's purpose with creation. In the light of such, Hefner consciously challenges Luther within the Lutheran tradition and calls the human being a "created co-creator". Luther had fiercely refused to accept the description of human beings as 'creators' so as not to blur the difference between creator and creature. His established wording to express the human role was "cooperatores non concreatores" (Luther 1912, p.857, 35). Hefner is primarily concerned with ethical considerations. He is not interested in putting the case for human arbitrariness within the evolutionary process. Rather, he intends to draw attention to the major significance of the human being's responsibility in light of his exponential position in the context of the evolutionary process. (cf. Hefner 1984; 1993; 1998).

Admittedly, Hefner's theological appropriation of evolutionary theory also draws on the described *eschatological turn* in the theology of creation, which also found its way into Anglo-American debates via the reception of Wolfhart Pannenberg and Jürgen Moltmann. Human responsibility with regard to non-human nature can only be described as 'creative' by presupposing and critically acknowledging that the end is still outstanding. Hefner underlines the semantic mark for this ontological and epistemological difference in two ways: by his tighter definition of 'created' and by the prefix 'co'. Thus Hefner is systematically rooted in the tradition of the transformation of creation theology in the modern age.

Nevertheless, a number of questions remain unanswered, because even Hefner still has mostly traditional genetic engineering in mind. The latter was, from both an ontological and epistemological perspective, mainly concerned with altering existing things. Hefner's approach, as a progressive adaptation of theological tradition, is criticized where conservative positions are advocated. At the same time, even progressive advocates will raise the question concerning the criteria according to which decisions are made. The fact that Hefner attributes "creativity" to the complexity of such technological decisions, which is not fully rationalizable, is basically not bold, but realistic. I will return to this point in conclusion.

But if it is correct that the dogmatic, and in particular the ethical debates, have been accentuated and aggravated by the emergence of synthetic biology, one would have to ask Hefner, as well as other advocates of the new theology of creation in the twentieth century, whether their affirmative approach to the technological modern age is also justified in this case.

4 Creativity and technology: Humans as co-creators

In the foregoing I have argued that synthetic biology raises new issues for the theological doctrine of creation. To illustrate this point I have pursued two lines of argumentation in parallel. First, I have shown how the theology of creation has changed in the modern age. The changes can be explained from the position of thought tradition, but also by the emergence of modern natural sciences. Traditional theological reservations have been transformed, or even abandoned. Second, I have underlined the need to negotiate certain fundamental concepts, which play a role in the theology of creation, in a new form as a result of synthetic biology. The conceptions of 'life' and of 'new matter', as well as ones of 'creativity' and 'technology' were important aspects. Evidently, the technical possibilities made use of in the field of synthetic biology also confront theology with new challenges. The reason is that the focus is no longer on changing existing things, but in fact, and radically so, on the idea of 'creating' 'new life'.

To address this issue I have shown that the generation of new matter is theologically linked to the eschaton and hence ascribed to God. New matter is immediately assigned an ultimate quality. At the same time, the new matter radically challenges what was previously valid. In a further step I have pointed out that the evolution of the theology of creation in the modern era, particularly in the twentieth century, is also heading in the same direction. This is because critical reflection on new technologies, particularly genetic engineering, has led to a broadly accepted *eschatological turn* in theology. The latter denotes an affirmative orientation regarding the natural-historical world, which has constituted itself as critical realism from an epistemical standpoint. Ontologically, it signifies linking the existent world to a fulfillment still to come. Used in an ethic-normative sense, this attitude implies a critical reservation towards the existent.

It is my contention that the issues associated with synthetic biology from a theological standpoint are not to be found primarily on an epistemological or ontological level, but are based on perplexities and shifts in the phenomenal sphere which confuse our everyday understanding of phenomena. I have pointed out that the assumption of a technical-rational process of generation contradicts the phenomenality of 'life' in a certain way. That applies in particular to 'new life', whose capacities and potentials are not completely assessable in all cases. The notion that 'new life' is produced by completely rational processes puts the involved scientists in a quasi-divine light. Not only do they seem to have power over what we are accustomed to perceiving as being genuinely stable and self-determined, but they are also burdened with heavy responsibility for the general ecological context. It is not by chance that the 'creative' aspect which is attendant on such processes has been moved theologically to that very ambivalent position that intuitively involves associations of omnipotence and omniresponsibility. Such intuitions have influenced how synthetic biology is treated in public and in the media.

I have shown that theology has a long affirmative tradition towards human beings' creative capacities. At the same time, it is befitting for such a tradition to stress that human activity is embedded in a context with other agents and agent-like entities. In recent theology of creation, reference to the independence of such other entities has been included in a temporal argument which also intends to include an ontological and epistemological element: reference to the correlation between the doctrine of creation and eschatology does not only throw a critical light on human capacities and possibilities in the light of an anticipated, but still outstanding ideal of fulfillment. Rather, it also heightens awareness of the implications of human activity and the limits of accountability. In this context, theological work is educational work. However, the critical reappraisal of the use of religious, in this case also Christian, semantics in public discourse has no destructive intention. On the contrary, it is a dual, affirmative and critical, reference to both the magnitude and obligatory character of scientific activity that is to be understood as a theological/

religious philosophical contribution to the debate on synthetic biology in the light of modern creation theology.

References

Altner, G. (2001). *Menschenwürde und biotechnischer Fortschritt im Horizont theologischer und sozialethischer Erwägungen.* München: Kaiser.
Améry, C. (1972). *Das Ende der Vorsehung. Die gnadenlosen Folgen des Christentums.* Reinbek: Rowolth.
Anselm, R. (2012). Schöpfung als Deutung der Lebenswirklichkeit. In: K. Schmid (ed.), *Schöpfung* (pp.225-294). Tübingen: Mohr Siebeck.
Barth, U. (1995). Abschied von der Kosmologie – Befreiung der Religion zu sich selbst. In: W. Gräb (ed.), *Urknall oder Schöpfung? Zum Dialog von Naturwissenschaft und Theologie* (pp.14-42). Gütersloh: Gütersloher Verlagshaus.
Bender, W., & Gerber, U. (1990). *Die selbstgestrickte Schöpfung. Gentechnologie: Was ist sie? Was kann sie? Was darf sie?* Stuttgart: Quell Verlag.
Blumenberg, H. (1997). *Paradigmen zu einer Metaphorologie.* Frankfurt: Suhrkamp.
Boldt, J., Müller, O., & Maio, G. (2009). *Synthetische Biologie. Eine ethisch-philosophische Analyse.* Bern: BBL.
Boldt, J., Müller, O., & Maio, G. (eds.) (2012a). *Leben schaffen? Ethische Reflexionen zur Synthetischen Biologie.* Münster: Mentis.
Boldt, J., Matern, H., Müller, O., Eichinger, T., & Ried, J. (2012b). Der Herstellungsbegriff in der Synthetischen Biologie. *Jahrbuch für Wissenschaft und Ethik, 17,* 89-116.
Dabrock, P. (2009). Playing God? Synthetic biology as a theological and ethical challenge. *Systems and Synthetic Biology, 3,* 47-54.
Dabrock, P., & Ried, J. (2010). Leben machen – Gott spielen? Theologische Anmerkungen zur Ethik der Synthetischen Biologie. *Die Politische Meinung, 487/55,* 37-41.
Dabrock, P., & Ried, J. (2011). Weder Schöpfer noch Plagiator. Theologisch---ethische Überlegungen zur Synthetischen Biologie zwischen Genesis und Hybris. *Zeitschrift für Evangelische Ethik, 55,* 179-191.
Dabrock, P., Bölker, M., Ried, J., & Braun, M. (eds.) (2011). *Was ist Leben – im Zeitalter seiner technischen Machbarkeit? Beiträge zur Ethik der Synthetischen Biologie.* Freiburg: Alber.
Danz, C. (2007). *Wirken Gottes. Zur Geschichte eines theologischen Grundbegriffs.* Neukirchen-Vluyn: Neukirchener.
Dohmen, K. (ed.). (1988). *Gentechnologie, die andere Schöpfung?* Stuttgart: Metzler.
Ebeling, G. (1989). *Lutherstudien. Band 2: Disputatio de homine 3. Teil.* Tübingen: Mohr Siebeck.
EKD (1991).Evangelische Kirche in Deutschland (Evangelical Church in Germany). *Einverständnis mit der Schöpfung. Ein Beitrag zur ethischen Urteilsbildung im Blick auf die Gentechnik und ihre Anwendung bei Mikroorganismen, Pflanzen und Tieren.* Gütersloh: Gütersloher Verlagshaus G. Mohn.
Fellmann, F. (2010). Leben. In: C. Bermes, U. Dierse (eds.), *Schlüsselbegriffe der Philosophie des 20. Jahrhunderts* (pp.189-206). Hamburg: Meiner.

Gansterer, G. (1997). *Die Ehrfurcht vor dem Leben*. Berlin: Lang.
Gerhardt, V. (2002). *Immanuel Kant. Vernunft und Leben*. Stuttgart: Reclam.
Graf, F.-W. (1990). Von der creatio ex nihilo zur 'Bewahrung der Schöpfung'. Dogmatische Erwägungen zur Frage nach einer möglichen ethischen Relevanz der Schöpfungslehre. *Zeitschrift für Theologie und Kirche, 87*, 206-223.
Härle, W. (2011). Luthers Rechtfertigungsverständnis. *Lateranum LXXVIII/1*, 125-138.
Haverkamp, A., & Mende, D. (eds.). (2009). *Metaphorologie. Zur Praxis von Theorie*. Frankfurt: Suhrkamp.
Hefner, P. (1984). The creation. In: C.E. Braaten, R.W. Jensons (eds.), *Christian Dogmatics. Vol. 1* (pp.269-357). Philadelphia.
Hefner, P. (1993). *The Human Factor. Evolution, Culture, and Religion*. Minneapolis: Fortress Press.
Hefner, P. (1998). Biocultural Evolution and the Created Co-Creator. In: T. Peters (ed.), *Science and Theology. The New Consonance* (pp.174-188). Boulder (Colorado): Westview.
Herms, E. (2011). Opus Dei gratiae. Cooperatio Dei et hominum. Luthers Darstellung seiner Rechtfertigungslehre in De servo arbitrio. *Lutherjahrbuch, 78*, 61-136.
Holm-Hadulla, R.M. (2011). *Kreativität zwischen Schöpfung und Zerstörung. Konzepte aus Kulturwissenschaften, Psychologie, Neurobiologie und ihre praktischen Anwendungen*. Göttingen: Vandenhoeck & Ruprecht.
Huber, W. (1990). *Konflikt und Konsens. Studien zur Ethik der Verantwortung*, München: Kaiser.
Huber, W. (2002). *Der gemachte Mensch*. Berlin: Wichern.
Hübner, J. (1986). *Die neue Verantwortung für das Leben. Ethik im Zeitalter von Gentechnologie und Umweltkrise*. München: Kaiser.
Hübner, J., & Schubert, H. v. (eds.). (1992). *Biotechnologie und evangelische Ethik*. Frankfurt: Campus.
Joest, W. (1967). *Ontologie der Person bei Luther*. Göttingen: Vandenhoeck & Ruprecht.
Käfer, A. (2004). Kant, Schleiermacher und die Welt als Kunstwerk Gottes. *Zeitschrift für Theologie und Kirche, 101*, 19-50.
Koch, T. (1991). *Das göttliche Gesetz der Natur. Zur Geschichte des neuzeitlichen Naturverständnisses und zu einer gegenwärtigen theologischen Lehre von der Schöpfung*. Zürich: Theologischer Verlag.
Körtner, U.H.J. (2005). *Lasset uns Menschen machen*. München: Beck.
Lakoff, G., & Johnson, M. (1980). *Metaphors we live by*. Chicago: University of Chicago Press.
Link, C. (1991). *Schöpfung. Schöpfungstheologie angesichts der Herausforderungen des 20. Jahrhunderts*. Gütersloh: Gütersloher Verlagshaus.
Link, C. (1997). *Gottesfrage und Schöpfungsglaube. Theologische Studien*. Neukirchen-Vluyn: Neukirchener Verlag.
Link, C. (2012). *Schöpfung*. Neukirchen-Vluyn: Neukirchener Verlag.
Losch, A. (2011). *Jenseits der Konflikte. Eine konstruktiv-kritische Auseinandersetzung von Theologie und Naturwissenschaft*. Göttingen: Vandenhoeck & Ruprecht.
Losch, A. (2014). Das Paradigma des Kritischen Realismus. In: C. Tapp, C. Breitsameter (eds.), *Theologie und Naturwissenschaften* (pp.69-94). Berlin: de Gruyter.
Luther, M. (1912). *D. Martin Luthers Werke. Kritische Gesamtausgabe* (1883-2009). 47. Band (WA 47). Weimar: Hermann Böhlaus Nachfolger.
Matern, H. (2015). Concursus sive cooperatio. Zur Transformation providenztheologischer Semantik in der Moderne. In: P. Dabrock, J. Ried, M. Braun (eds.), *Bio-Objekte. Theologische Grundlagen für eine Ethik der Biowissenschaften*. Stuttgart: Kohlhammer (im Druck).

Meadows, D.L., Meadows, D.H., Zahn, E., & Milling, P. (1972). *Die Grenzen des Wachstums. Bericht des Club of Rome zur Lage der Menschheit.* Stuttgart: Deutsche Verlags-Anstalt.
Moltmann, J. (1987). *Gott in der Schöpfung. Ökologische Schöpfungslehre.* Gütersloh: Gütersloher Verlagshaus.
Nordmann, A. (2007). Renaissance der Allianztechnik? Neue Technologien für alte Utopien. In: B. Sitter-Liver (ed.), *Utopie heute: Zur aktuellen Bedeutung, Funktion und Kritik des utopischen Denkens und Vorstellens. Vol. 1.* (pp.261-278). Fribourg: Academic Press Fribourg.
Pannenberg, W. (1991). *Systematische Theologie. Band 2.* Göttingen: Vandenhoeck & Ruprecht.
Ried, J. (2011). Der Mensch als created co-creator. Anmerkungen zu einem theologisch-anthropologischen Konzept der kreatürlichen Verantwortlichkeit. In: P. Dabrock, M. Bölker, J. Ried, M. Braun (eds.), *Was ist Leben – im Zeitalter seiner technischen Machbarkeit? Beiträge zur Ethik der Synthetischen Biologie* (pp.103-117). Freiburg: Alber.
Ried J., Braun M., Dabrock P. (2011). Unbehagen und kulturelles Gedächtnis. Beobachtungen zur gesellschaftlichen Deutungsunsicherheit gegenüber Synthetischer Biologie. In: P. Dabrock, M. Bölker, J. Ried, M. Braun (eds.), *Was ist Leben – im Zeitalter seiner technischen Machbarkeit? Beiträge zur Ethik der Synthetischen Biologie* (pp.345-367). Freiburg: Alber.
Sass, H. v. (2013). Wahrhaft Neues? Eine einleitende Erinnerung. In: H. von Sass (ed.), *Wahrhaft Neues. Zu einer Grundfigur christlichen Glaubens* (pp.9-50). ThLZ.F 28. Leipzig: Evangelische Verlagsanstalt.
Schmid, K. (2012). *Schöpfung.* Tübingen: Mohr Siebeck.
Schrauwers, A., & Poolman, B. (2013). *Synthetische Biologie. Der Mensch als Schöpfer?* Berlin: Springer Spektrum.
Schummer, J. (2011). *Das Gotteshandwerk – Die künstliche Herstellung von Leben im Labor.* Berlin: Suhrkamp.
Schweitzer, A. (1966). *Die Lehre von der Ehrfurcht vor dem Leben.* München: Beck.
Tauber C. (2001). 'Uomo universal' oder 'Uomo virtuoso'? Zum Menschenbild der Renaissance. In: D. Ansorge, D. Geuenich, W. Loth (eds.), *Wegmarken europäischer Zivilisation* (pp.178-203). Göttingen: Wallstein.
Thomas, G. (2013). Emergenz oder Intervention? Konstellationen der schöpferischen Treue Gottes in Auseinandersetzung mit einem theologischen Naturalismus. In: H. von Sass (ed.), *Wahrhaft Neues. Zu einer Grundfigur christlichen Glaubens* (pp.151-190). ThLZ.F 28. Leipzig: Evangelische Verlagsanstalt.
Thompson, M. (2008). *Life And Action. Elementary Structures of Practice and Practical Thought.* Boston: Harvard University Press.
Wehowsky, S. (1985). *Schöpfer Mensch? Gen-Technik, Verantwortung und unsere Zukunft.* Gütersloh: Gütersloher Verlagshaus G. Mohn.
Welker, M. (1995). Was ist Schöpfung? Genesis 1 und 2 neu gelesen. In: M. Welker, *Schöpfung und Wirklichkeit.* (pp.15-41). Neukirchen-Vluyn: Neukirchener Verlag.
Weth, R. (2004). *Der machbare Mensch.* Neukirchen-Vluyn: Neukirchener Verlag.
White, L. (1967). The Historical Roots of Our Ecologic Crisis. *Science*, 155(3767), 1203-1207.
White, L. (1973). Continuing the Conversation. In: I.G. Barbour (ed.), *Western Man and Environmental Ethics – Attitudes Toward Nature and Technology* (pp.55-64). Reading (Mass.): Addison-Wesley Publishing Company.
Witt, E. (2012): *Konzepte und Konstruktionen des Lebenden. Philosophische und biologische Aspekte einer künstlichen Herstellung von Mikroorganismen.* Freiburg: Alber.
Wright, A. (2013). *Christianity and critical realism.* London: Routledge.
Zenger, E. et al. (1995). *Einleitung in das Alte Testament.* Stuttgart: Kohlhammer.

The moral economy of synthetic biology

Bernadette Bensaude Vincent

1   Moral economy as a heuristic tool

In merging engineering and biology and electing design as the main focus of research, synthetic biology is likely to bring deep changes in the set of norms and values that used to rule scientific research. Is it generating a new profile of biologist? Based on Lorraine Daston's concept of the moral economy of science, this paper explores the affects, values and norms attached to the research activities of synthetic biologists. It endeavors to disentangle the values and norms underlying their writings and interviews. More broadly, the purpose is to follow the genesis of a moral economy of science based on a specific case.

Beginning with a brief definition of the concept of moral economy, the paper proceeds to discuss whether synthetic biology subverts the values and norms that compose the moral economy of science. Finally, it will try to identify the distinctive emotional component of the new moral economy that rules synthetic biology. Daston's concept of the moral economy of science, developed in a landmark paper (Daston 1995) is not to be confounded with the psychological motivations of individual scientists, since it refers to the values and norms underlying the daily work of a collective of researchers (close to Ludwik Fleck's *Denkkollektiv*). It is not, however, to be identified with a universal 'scientific spirit', because it is not immutable and depends on the context. It also differs from ideology since it is embedded in scientific practice. The norms and values that underlie scientific activity are intimately connected to the collective practices developed in a research field. At the same time, they are related to the broader values of society and culture.

Daston's use of the term 'economy' comes close to the older sense of a system of domestic rules and has nothing to do with questions of money, market or labor. Unlike the concept of moral economy forged by Edward Thompson (1971) to account for workers' hunger riots,[1] Daston does not claim to provide any causal account of the behavior of scientists. She simply assumes that a consistent and global set of norms and values comes out of various individual endeavors or commitments. She insists on the contingency of the scientists' moral economy, and she also emphasizes the role of emotions in science and their embedment in a cultural context.

> A moral economy is a balanced system of emotional forces, with equilibrium points and constraints. Although it is a contingent malleable thing of no necessity, a moral economy has a certain logic to its compositions and operations. Not all conceivable combinations of affects and values are in fact possible (Daston 1995, p. 2).

In addition, Daston confers a certain degree of stability to this set of values while conceding that it is likely to evolve over time and differ from one cultural milieu to another.

In her effort to grasp the moral components of modern science Daston developed three examples of epistemic ideals – quantification, empiricism, and objectivity – to show how scientists have attempted to put them into practice over the past centuries by developing a number of 'epistemic virtues'.

Quantification, she argued, did not prompt a search for exact correspondence between mathematical formulations and phenomena. More than exactitude it valued accuracy, as instantiated in Leibniz's *characteristic universalis*. And quantification mainly served to silence dissent and close controversies.

The empiricist ideal, developed through the codification of experimental practices in the academic world, relied on witnesses or, more precisely, on the trust in witnesses and in the researcher more than on effective replication of experiments.

The form of objectivity presented in her paper is the ideal of a view from nowhere. This non-perspective view of the world requiring the anonymity of scientists favored practices of sharing and openness.

What happens to such epistemic ideals and their associated virtues in a milieu such as a research setting oriented toward the design of genetic circuits, minimal cells or synthetic bacteria? Are they still ruling the community or are they competing with other values and norms?[2]

---

1   Edward Palmer Thompson forged the concept in 1971 to provide an ethnographical, not strictly Marxist account for the hunger riots of workers. James Scott used it in the 1990s to emphasize that peasant revolts and protests were inspired by a system of values such as social justice (cf. Fassin 2009).
2   In this paper the basic distinction made by moral philosophers between *values*, referring to judgments about what is right and wrong, and *norms*, ruling what to do or not to

The moral economy of synthetic biology 89

Furthermore, many scholars have noted that synthetic biology is an umbrella term for diverse research agendas (O'Malley et al. 2008; Deplazes 2009), which are based on quite different epistemic models (Bensaude Vincent 2013). And Sacha Loeve (in this volume) argues that such diversity goes far beyond the four research programs usually identified –biobricks, minimal cell, protocells, and xenobiology – and that it has not been reduced over the past decade. If epistemic pluralism is a distinctive feature of synthetic biology, is it possible that its practitioners share a common set of values and affects?

2 Subverting values and norms

Looking at how synthetic biologists describe their research practice it seems that the engineering ideal has taken over the epistemic ideal. The strong claim made by Robert Carlson (2010) that "Biology is technology" suggests a disruption of the conventional ideals of life scientists who aim at understanding life.

In particular, if the main goal of synthetic biology is engineering or redesigning life, it seems that objectivity is no longer to be pursued. Truth – in the sense of correspondence between the mind and reality – is no longer the ultimate value. Neither the eighteenth-century notion of "truth-to-nature" nor the nineteenth-century "mechanical objectivity" (Daston and Galison 2007) can be a model as long as synthetic biologists are not primarily interested in reaching an agreement between theory and reality. Instead of providing an 'objective' image of the molecular mechanisms at work in living cells, they seek to remake them or to make unnatural microorganisms.

To be sure, the desire of creating things that do not exist does not preclude all ambitions to gain objective knowledge about living organisms. Quite the contrary. Synthetic biologists, as Maureen O'Malley argues, distinguish themselves by their "focus on *making* as true knowledge" (O'Malley, 2009, p.385). They claim they are "knowing through making." David Sprinzak and Michael Elowitz, both involved in the construction of genetic circuits, express their credo in the role of 'making' by resorting to the conventional metaphor of the clock:

> By taking apart an old clock, you could probably come up with a pretty good guess at how it works. But a more concrete understanding of the clock mechanism might be obtained by designing and building one's own clock out of similar parts. Contemporary biology presents us with similar reverse-engineering problems. It's reverse

> do, can be overlooked. The evaluative and the prescriptive converge to determine the choices and actions of scientific researchers.

analysis although modelling and simulating are used beforehand (Sprinzak and Elowitz 2005, p.443).

Synthetic biologists consequently like to present systems biology and synthetic biology as two complementary steps in a cognitive enterprise. Ron Weiss, who is engaged in the design of mammalian cells, clearly states:

> Synthetic biology complements systems biology: the former is based on forward engineering and the latter on reverse engineering. For instance, insight gained from systems biology investigations of natural processes leads to improved designs of synthetic systems, and the creation of small artificial networks helps to analyse hypotheses on the function of natural ones (Church et al. 2014, p.289).

Even when the design practices of synthetic biologists are driven by a curiosity for natural organisms, their cognitive activities shift from a concern about reality toward a concern about potentiality: exploring all possible combinations of bases to make exotic nucleic acids, or amino acids, exploring the potentials of cells viewed as kinds of atomic architects... the target of research is the realm of the possible.

In this systematic exploration of potentialities, the terms 'making' and 'understanding' lose their usual meanings. 'Making' does not mean designing and crafting something. It primarily refers to *in silico* modeling and automated high-throughput analytical and synthetic methods. On the other hand, the term 'understanding' no longer means providing a rational explanation of natural mechanisms and causalities. As Alfred Nordmann wrote: 'Understanding' in this case does not refer to the mind's intellectual faculty. It's a more pragmatic notion related to the action of monitoring the behaviour and the performances or research objects (Nordmann 2015).

Knowing through making requires other 'virtues' than the modesty and anonymity of the scientists struggling for objectivity. It requires control of the engineering process. The frequency of this term in interviews with synthetic biologists suggests an obsession with control: control of the instructions or program, control over the circuitry for turning genes on and off, control over hundreds of genes simultaneously, control over intercellular communication.[3] More generally, synthetic biologists seem to be revitalizing the modern ideal of men shaping and controlling the world out there according to their intents and projects.

Although objectivity is no longer their ideal, synthetic biologists promote sharing or communalism as a major virtue. While unrestricted exchange of information and materials was justified in Robert Merton's famous scientific ethos of science as a

---

3   See, for instance, the interviews with G.M. Church, M B. Elowitz, C. D. Smolke, C.A. Voigt and R. Weiss (2014).

precondition to reach objective truth through maximized interactions, the BioBrick community promotes the ideal of openness based on the principle of 'get some, give some'. The Registry of Standardized Bioparts benefits its users who, in return, should improve the community resource (Hilgartner 2012). Justifications refer to the empowerment of creative talents and the take-off of a new bioeconomy based on a critical mass of materials and data (Carlson 2010). The epistemic virtue of communalism thus serves a new ideal. Efficiency displaces objectivity as ultimate value.

The culture of experiment developed in synthetic biology requires other virtues than those of seventeenth-century experimental philosophers. Empiricism is no longer a question of constructing and stabilizing 'hard facts' capable of withstanding all objections and resisting all attempts at falsification. As facticity required confrontation and refutation, polemics has been considered a major attribute of Gaston Bachelard's "scientific spirit", and Merton's scientific ethos included organized skepticism. Synthetic biologists, by contrast, seem to be little inclined towards confrontation and do not praise a critical spirit. Although the research community is made up of a variety of groups with quite distinct research agendas there is no visible rivalry between the tribes. Tolerance seems to be a major virtue.

Synthetic biologists are not fighting for truth, they are content with the administration of proofs of concepts. Their objective is to demonstrate that an idea, or a principle, has the potential to be used for producing new functional devices. For instance, the construction of a genetic toggle switch in E. coli (Gardner et al. 2000) and the design of an oscillatory repressilator (Elowitz and Leibler 2000) were two landmark episodes in the beginnings of synthetic biology. And it does not matter that, a decade later, those concepts have not been turned into technological applications. In particular, Elowitz's and Leibler's repressilator never really worked: only 40% of cells oscillated, and the oscillation stopped after ten hours. In the traditional moral economy of science such disappointing results would have generated distrust. Colleagues would have raised doubts and criticism about the theoretical assumptions that inspired this proof of concept. A controversy would presumably have ensued, with a battle of rival experiments attempting to end it. But nothing like that happened: no objections, no controversy, no distrust, no discredit. The failure in the design of a synthetic circuit did not raise any skepticism about the bioparts agenda, which inspired the design, because this agenda is not conceived of as a hypothesis to be tested by experiments.

The Baconian ideal of crucial experiment gives way to empiricism by a trial-and-error mode. The crucial virtue is learning from failure: Elowitz and his collaborators consider how to integrate the obstacles into the design of synthetic genetic circuits. They try to integrate noise into the circuit, to take into account the intrinsic stochasticity of the cell and the non-linearity of effects. Such attitude could have challenged the ideal of control and design, since that makes the output partly unpredictable. But it did not:

> The last decade has shown how even our first steps toward building and analyzing synthetic circuits can identify fundamental biological design principles and can produce useful new understanding. Future progress will require work across a range of synthetic levels [...] from rewiring to building autonomous and integrated circuits de novo. Going forward, we anticipate that synthetic biology will become one of the primary tools we use to understand, control, imagine, and create biological systems (Nandagopal and Elowitz 2011, p.1248).

Thus the experimental proofs of concept are immune to reality tests. They open up dazzling and desirable futures, which become the unique reference for securing trust and creating matters of concern. Reality is displaced by a horizon of possibilities.

Finally, quantification may be the unique ideal that has not been subverted in synthetic biology. For many biologists quantitative models are key to successfully designing bio-systems. They seek to make biology as quantitative and rigorous as physics and engineering (Calvert and Fujimura 2009). Carlson, for instance, insists that for engineering biology it is essential to shift from natural language to empirical model and then to quantitative model (Carlson 2010, pp.40-41). In his view, quantitative models alone provide predictions about the outcome and allow control. In that case, quantification basically refers to computing, on the basis of molecular data, in the same way that Leibniz's *characteristic universalis* was supposed to compute all the possible worlds.

Quantification in modern science, as described by Daston, secured communicability while at the same time creating a distance between science and nature. Similarly, the ideal of quantification in synthetic biology helps develop consensus in a community. In particular, precision measurement is required for the characterization of bioparts. The quantification of all characteristics is a necessary precondition to achieve their standardization, which is one of the basic rules of engineering biology formulated by Drew Endy (2005). In this case, quantification is first and foremost an effort to reduce complexity. Abstraction and decoupling, the other two rules, clearly indicate that reducing complexity is also an epistemic virtue: "No biological engineer will succeed in building a system *de novo* until most of that complexity is stripped away, leaving only the barest essentials" (Carlson 2010, p.6). Understanding complexity may still be part of the horizon, but compared to the engineering ideals this objective does not loom large.

In promoting the design of biosystems as its ultimate goal, synthetic biology partly undermines the moral economy of science. Objectivity has been displaced by the ideals of control and efficiency; empiricism is more like trial than tribunal; and the ultimate value is to display potentialities rather than mirror reality. Quantification alone retains its virtues of sociability and emancipation from the constraint of reality.

3 Playfulness as a virtue

If synthetic biologists subvert some of the values and norms that used to rule scientific activities, do they promote an alternative moral economy? I will argue that they radically change the emotional components of the moral economy of science: from the sober self-dedication of disinterested scientists to a playful activity of creating toys.

To be sure, synthetic biology is advertised as an earnest business driven by 'serious' concerns regarding health, environment, and competitiveness. These societal commitments are always mentioned to legitimize research policy and investments in the field. They create an aura of responsible research and innovation.

Science policy discourses are, however, not the best source material to unveil the moral economy of a science. It is more through the scientists' writings and interviews that one can investigate their moral commitments, worldviews, and values. In the case of synthetic biology there are plenty of popular essays written by star scientists, regular interviews and review articles assessing the state of the art.

Remarkably, just a few scientists seem to drive the field and shape its moral economy. A recent scientometric study shows that since 2004 only eight authors – George Church, James J. Collins, Drew Endy, Jay Keasling, Wendell Lim, Christina D. Smolke, Christopher A. Voigt, and Ron Weiss. – have furnished more than 60% of citations (Raimbault and Joly, in press). These leading figures are all working in US laboratories. They are regularly interviewed in periodicals such as *Nature* and *Science*. They act as institutional entrepreneurs, assuming public relations duties and securing the visibility of the field in international media.

These "sociotechnical vanguards" (Hilgartner 2015) distinguish their field as an exciting adventure open to laypeople and attracting young and imaginative talents. They claim that engineering life is cool, pleasant, and entertaining. Like the geeks generating software and added value while pursuing their addiction to video games, synthetic biologists claim to advance science while playing a collective game. They frame the community on the model of team sports centers. The iGEM (International Genetically Engineered Machine) competition organizes contests between teams from all around the world. It is dedicated to advancing synthetic biology in a playful and educative manner with an inspiration derived from the Boy Scout movement.

Developing a community of practitioners on the basis of teamwork and playful activities is one of the major impacts of this successful institution.

To be cool and playful, engineering life has to be perceived as being easily accessible to laypeople. Drew Endy is especially committed to making biology easy to engineer. In an interview in 2008, he claimed that just as one does not need expert knowledge about pigments and tools to become a designer in graphic arts, one can design life without acquiring expert knowledge. And in conclusion he stated: "How to make biology easy to engineer? I don't want to talk about it, I want to do it" (Brockman 2008). On a similar note, Carlson claims that designing life is simple, easy, and accessible to laypeople.

> Many standard laboratory techniques that once required a doctorate's worth of knowledge and experience to execute correctly are now used by undergraduates in a research setting with kits containing color-coded bottles of reagents. The recipes are easy to follow. This change in technology represents a democratization of sorts, and it illustrates the likely changes in labor structure that will accompany the blossoming of biological technology (Carlson 2001, p.3).

He is a staunch advocate of garage biology. In his view, "an economy based on the rational engineering of biological systems is likely to require a library of composable parts with defined behaviors" (ibid., p.174). On the model of the diffusion of hacker culture around the world, amateur biologists are encouraged to set up labs in their garages or in community centers. Thanks to the Registry of Standardized Bioparts and the availability of PCR (Polymer Chain Reaction) and electrophoresis equipment, garage biology is already well underway. Whether DIY biologists are motivated by a simple desire to learn or by political inclinations towards the democratization of knowledge, they all insist on the fun of it.

Even the serious concerns raised by DIY biologists about biosecurity are addressed in a playful manner. In the SYNBERC consortium, the ethicists in charge of practice and policy issues at the Human Practices Thrust have been replaced by Drew Endy, who arranged a collaboration with the police. F.B.I. agents are now involved in the playful activities of iGEM students and DIY biologists (Angeli Aguiton and Tocchetti 2015).

In the symbolic order, synthetic biologists attach this emotional charge of playfulness to their research practice by using two favorite metaphors:[4] playing LEGO and rewriting the code of life.

---

[4] The symbolic order is a conceptual network through which humans perceive the natural world at a collective, societal level as opposed to individual psychological mental models.

The LEGO metaphor is not specific. It has been frequently used in nanotechnology since it is a powerful image to convey the view of design from the bottom up, atom-by-atom, or brick-by-brick advertised by Eric Drexler (1986) and the Center for Responsible Nanotechnology. It has survived the criticisms of scientists (including Nobel Prize winner Richard Smalley) who objected that chemistry at the nanoscale relied on solvents and low binding energies rather than on rigid blocks thoroughly clamped together, or that there is a special physics in the nanoworld (Jones 2004). The LEGO metaphor seems unaffected by criticisms. It is still being used not only as an educational tool in nanotechnology classes or summer schools but for research purposes as well. Carlson considers this resilient metaphor to be totally appropriate:

> Building with LEGOS is an excellent metaphor for future building with biology. The utility and unifying feature of LEGOS, Tinkertoys, Erector Sets, Zoob, or Tente is that the pieces fit together in very understandable and defined ways. This is not to say they are inflexible – with a little imagination extraordinary structures can be built from LEGOS and the other systems of parts. But it is easy to see how two bricks (or any of the other newfangled shapes) can fit together just by looking at them (Carlson 2010, p.8).

Carlson tries to rescue the metaphor from the objections raised by physicists and chemists. While he is aware that biobricks are not rigid like LEGO blocks, he thinks the metaphor adequately conveys the bioparts agenda, i.e., the view of bricks arranged into modules and then systems. It helps spread the view of a system in which every elementary unit has its purpose. Carlson does, however, concede that we do not know all the details and that the LEGO metaphor has its limitations:

> The mechanical intuition we have for gears, motors, beams and cantilevers cannot be simply transferred to molecules. The molecular scale of the components also means they are very hard to interact with directly [...]. Designing biological systems just isn't like imagining a new LEGO structure. But even given our inadequate knowledge of biological details, a design philosophy based on composable parts still has utility (Carlson 2010, p.37).

The apparent contradiction between the two above quotations can be resolved if we assume that the LEGO metaphor is not meant to convey a mechanical view of life. It has no representational function and is rather used in a performative way to promote the bioparts agenda. Carlson is not a naïve reductionist ignoring all the complexity of biological systems. But the message he wants to convey with the

---

The potential of metaphors for generating new meanings and values in science has been acknowledged (Maasen and Weingart 2000).

LEGO metaphor is that the future of synthetic biology requires a methodological reductionism. It's a kind of make-believe game. The LEGO blocks suggest that it is possible to design and standardize interchangeable bioparts, and to proceed from the simple to the complex by using quantitative and predictive models. In other terms, the metaphor makes the case for a strategy of design mobilizing creative skills close to those developed through children's games such as Bob the Builder or LEGO sets.

The LEGO metaphor thus makes sense to surround synthetic biology with a light-hearted ambiance of entertainment and conviviality. It is an integral part of a process of gamification strongly encouraged by the synthetic biology community and supported by non-profit and profit groups. Gamifying synthetic biology has been a major concern since the SB6.0 Conference in 2013. One can "plug and play with biology" with the Syn Mod App from App Store, while the DIY association Leukippos gives everyone the opportunity to collaborate on synthetic biology and form a "biology lab in the cloud."

With regard to the standard distinction between strategy and tactics, tactical games seem to be the prevailing model in the gamification of synthetic biology. While strategic games mobilize a broad plan based on high awareness of the situation and the order of things, tactical games rely on isolated actions. Synthetic biologists similarly take advantage of local opportunities to create objects and move forward.

The metaphor of the genetic code that proved extremely heuristic in twentieth-century biology (Fox Keller 1995; 2002) is still ubiquitous in synthetic biology. It is used for branding its agenda as one aimed at reading and rewriting DNA and genomes. Craig Venter has spread his ambition of "rewriting the software of life" in all media. He consciously enriched the metaphor in suggesting that the human genome project was a shift from analog to digital computation:

> For the past 15 years at ever faster rates we have been digitising biology. [...] We and others have been working for the past several years on the ability to go from reading the genetic code to learning how to write it. It is now possible to design in the computer and then chemically make in the laboratory, very large molecules (Venter 2007).

Tom Knight added a flavor of urgency in asserting: "The genetic code is 3.6 billion years old. It's time for a rewrite" (cf. Silver 2007). Synthetic biologists are fond of the informational metaphor for two obvious reasons: on the one hand, rewriting, being a more radical enterprise than 'gene modification', creates a semantic distance between synthetic biology and genetic engineering. Such a distance is crucial for the public acceptance of synthetic biology, especially in Europe where GM crops are rejected. On the other hand, rewriting the code of life is an act of creation. In France, the DNA rewriting exercise has been conceived of as a true literary game.

French geneticist Philippe Marlière took inspiration from a surrealist game jointly designed in the mid-twentieth century by writers and mathematicians. The *Ouvroir de Littérature Potentielle* (opening the potentials of literature), abbreviated as OULIPO, served as a model for the concept of OUGEPO (*Ouvroir de Génétique Potentielle*), a biological game based on the combinatorial principle of genetic sequences meant to create exotic amino acids.

In any case, the emphasis on creation allows synthetic biologists to play God. They like to make believe that they hold life in their hands, and that humans become creators of life (Boldt and Müller 2008). George Church, for instance, did not hesitate to pen a popular book entitled *Regenesis*. His grand narrative of redesigning life, from its dawn on Earth to future post-humans immune to all viruses, relies on the conviction that the genetic code is a universal machine enabling synthetic biologists to "reinvent nature and ourselves":

> Biological organisms could be viewed as a kind of high technology, as nature's own versatile engines of creation. Just as computers were universal machines in the sense that given the appropriate programming they could simulate the activities of any other machine, so biological organisms approached the condition of being universal constructors in the sense that with appropriate changes to their genetic programming, they could be made to produce practically any imaginable artefact. A living organism, after all, was a ready-made, prefabricated production system that, like a computer, was governed by a program, its genome. Synthetic biology and synthetic genomics, the large-scale remaking of a genome, were attempts to capitalize on the facts that biological organisms are programmable manufacturing systems, and by making small changes in their genetic software a bioengineer can effect big changes in their output (Church and Regis 2012, p.4).

Indeed, "Playing God" is not to be taken at face value. Although James Watson allegedly declared: "If scientists don't play God, who else is going to?" biologists such as Venter and Church are well aware that they cannot assume the role of God. They do not really create life, since they have to rely on living cells to express the synthetic genomes. Synthetic biologists like to play God for the sake of playfulness. There is no religious aura in their broad claims, which do not cause uproar in religious milieus. While theologians have raised serious objections about stem cell research, they are not concerned about synthetic biology (Belt 2009). "Playing God" is just a rhetorical stance used for attracting public attention on synthetic biology. It conveys less a metaphysical *hubris* and more a sportive activity aimed at spectacular records. Just like the champions of extreme sports, synthetic biologists display spectacular tours de force in the media and feature as the stars of new human adventures. Playing God is just another manner of gamifying synthetic biology by playing with the limits of nature. Synthetic biology is experienced as an amusing pursuit devoid of serious concern about security, ethics or religion. One would never accuse synthetic biologists of blasphemy. Humor is a distinctive trait of their activity and sociability. Endy, for instance, very much appreciates the cartoons meant for mocking his research management (Angeli Aguiton 2014).

To play God, synthetic biologists have to show a marked disdain for the technical aspects of the design of life. It's just programming! When Craig Venter, for instance, presented the synthetic bacterium *Mycoplasma Laboratorium* in 2010 as "the first self-replicating species we've had on the planet whose parent is a computer" (Wade 2010), he simply sidelined the hard and painstaking years of efforts of dozens of technicians and highly skilled collaborators from his Center who had done all the biochemical work to insert the synthetic genome into a living cell and get it to express the imported genome. The metaphor of the program encourages active ignorance, a form of 'agnotology', regarding the 'wetware', or the technical and material aspects of the design of living organisms. The dematerialization of life is part and parcel of the gamification of synthetic biology, as it downplays the cost and difficulties of the design. Unexpectedly, most of the objects designed by synthetic biologists up to now are 'toy' applications or gadgets. Miles away from the constraints of 'real-world', they furbish a dream world and foster a massive depoliticization of scientists.

In less than a decade, a dozen of vanguards active in academia and in the media have been able to bend the moral economy of science in new directions. Not only did they undermine some of the pillars of the moral economy forged over the past centuries, but they also introduced a new register of affects – easiness, playfulness, entertainment, game – which were usually kept out of the sphere of knowledge production and confined to science popularization or recreational science. Undoubtedly, this change reflects the fact that synthetic biology is part of public culture and that the norms and values developed by hackers and geeks are all-pervasive.

There is no evidence that the emerging moral economy has the potential to cement and stabilize an international community of synthetic biologists and to overcome the diversity of their research agendas. Not all active researchers in the field share those values and affects or feel at home in this moral economy. A number of European researchers have expressed quite different values and emotions in their response to a questionnaire about their worldviews and values (Loeve 2014). They tend to praise complexity more than easiness and playfulness. They sometimes express concern about the boundary between the natural and the artificial, the living and the non-living worlds/entities. Yet there is still no sign of active opposition on the part of European synthetic biologists to the values and affects developed by the promoters of the field in the USA.

This analysis of the moral economy of synthetic biology will hopefully provide a basis for a constructive critique of synthetic biology. Highlighting the norms and values that have shaped the field over the past decade may invite a number of synthetic biologists to express alternative commitments, and, on the other hand, it may serve as a basis for debates in the public arena about the norms and values that 'should' rule synthetic biology.

References

Angeli Aguiton, S. (2014). *La démocratie des chimères: gouvernement des risques et des critiques de la biologie synthétique, en France et aux Etats-Unis.* Thesis/Dissertation. Institut d'études politiques Sciences-Po Paris.
Angeli Aguiton, S., & Tocchetti, S. (2015) Is a FBI agent a DIY biologist as any other? *Science Technology and Human Values* (in print).
Belt, H. v. d. (2009). Playing God in Frankenstein's footsteps: Synthetic biology and the meaning of life. *Nanoethics, 3*(3), 257–268.
Bensaude Vincent, B. (2013). Discipline-building in synthetic biology. *Studies in History and Philosophy of Science Part C: Studies in History and Philosophy of Biological and Biomedical Sciences, 44*(2), 122-129.
Boldt, J., & Müller, O. (2008). Newtons of the leaves of grass. *Nature Biotechnology, 26*(4), 387-389.
Brockman, J. (2008). Engineering Biology: A Talk with Drew Endy. *Edge: The Third Culture.* http://www.edge.org/3rd_culture/endy08/endy08_index.html. Accessed: 15 May 2015.
Calvert, J., & Fujimura, J. H. (2009). Calculating life? *EMBO Reports, 10*(1S), S46-S49.
Carlson, R. (2001). *Biological Technology in 2050.* http://www.synthesis.cc/Biol_Tech_2050.pdf. Accessed: 27 January 2015.
Carlson, R.H. (2010). *Biology is Technology: The Promise, Peril, and New Business of Engineering Life* (Vol. 2010). Cambridge, MA: Harvard University Press.
Church, G., & Regis, E. (2012). *Regenesis: How Synthetic Biology Will Reinvent Nature & Ourselves.* New York: Basic Books.
Church, G. M., Elowitz, M. B., Smolke, C. D., Voigt, C. A., & Weiss, R. (2014). Realizing the potential of synthetic biology. *Nature Reviews Molecular Cell Biology, 15*(4), 289-294.
Daston, L. (1995). The moral economy of science. *Osiris, 10,* 2-24.
Daston, L., & Galison, P. (2007). *Objectivity.* New York: Zone Books.
Deplazes, A. (2009). Piecing together a puzzle. *EMBO Reports, 10,* 428-432.
Drexler, E. (1986). *Engines of Creation.* New York: Anchor Books.
Endy, D. (2005). Foundations for engineering biology. *Nature, 438,* 449-453.
Fassin, D. (2009). Les économies morales revisitées. *Annales. Histoire, Sciences sociales, 64*(6), 1237-1266.
Fox Keller, E. (1995). *Refiguring Life: Metaphors of Twentieth-Century Biology.* New York: Columbia University Press.

Fox Keller, E. (2002). *Making Sense of Life: Explaining Biological Development with Models, Metaphors, and Machines.* Harvard: Harvard University Press.

Hilgartner, S. (2012). Novel Constitutions? New Regimes of Openness in Synthetic Biology. *BioSocieties, 7*, 188-207.

Hilgartner, S. (2015). Capturing the Imaginary: Vanguards, Visions and the Synthetic Biology Revolution. In: S. Hilgartner, C. Miller, R. Hagendijk (eds.), *Science & Democracy: Knowledge as Wealth and Power in the Biosciences and Beyond* (pp.33-55). New York, Abingdon: Routledge.

Jones, R. (2004). *Soft Machines.* Oxford, New York: Oxford University Press.

Loeve, S. (2014). Summary Report of the Paris Workshop *Worldviews and Values in Synthetic Biology.* Synenergene. http://synenergene.eu/resource/summary-report-paris-workshop-worldviews-and-values-synthetic-biology. Accessed: 5 May 2015.

Maasen, S., & Weingart, P. (2000). *Metaphors and the Dynamics of Knowledge.* London: Routledge.

Nandagopal, N., & Elowitz, M. B. (2011). Synthetic biology: integrated gene circuits. *Science, 333*(6047), 1244-1248.

Nordmann, A. (2015). Synthetic biology at the limits of science. In: B.M. Giese, C. Pade, H. Wigger, A. von Gleich (eds.), *Synthetic Biology: Character and Impact* (pp.31-58). Springer International Publishing Switzerland.

O'Malley, M.A. (2009). Making knowledge in synthetic biology: Design meets kludge. *Biological Theory, 4*, 378-389. http://www.maureenomalley.org/downloads/OMalley-BioTheo-Design-Kludge.pdf. Accessed: 15 May 2015.

O'Malley, M.A., Powell, A., Davies, J.F., & Calvert, J. (2008). Knowledge-making distinctions in synthetic biology. *BioEssays, 30*(1), 57-65.

Raimbault, B., & Joly, P.B. (forthcoming). On the emergence of techno-scientific fields: The case of Synthetic Biology...

Salter, B., & Salter, C. (2010). Governing innovation in the biomedicine knowledge economy: stem cell science in the USA. *Science And Public Policy, 37*(2), 87-100.

Silver, L. (2007). Scientists push the boundaries of human life. *Newsweek*, 3 June 2007. http://www.newsweek.com/scientists-push-boundaries-human-life-101723. Accessed: 15 May 2015.

Smalley, R. (2001). Of Chemistry, Love and Nanobots: How soon will we see the nanometer-scale robots envisaged by K. Eric Drexler and other molecular nanotechnologists? The simple answer is never. *Scientific American*, Sept. 2001, 76-77.

Sprinzack, D., & Elowitz, M.B. (2005). Reconstruction of genetic circuits. *Nature, 348*, 443-448.

Thompson, E. P. (1971). The moral economy of the English crowd in the eighteenth century. *Past & Present, 50*, 76-136.

Venter, C. (2007). A DNA-Driven World: The 32nd Richard Dimbleby Lecture. BBC One, 4 December 2007. www.bbc.co.uk/pressoffice/pressreleases/stories/2007/12_december/05/dimbleby.shtml. Accessed: 26 January 2015.

Wade, N. (2010). Researchers Say They Created a 'Synthetic Cell'. *The New York Times*, 20 May 2010. http://www.nytimes.com/2010/05/21/science/21cell.html. Accessed: 15 May 2015.

Evaluating biological artifacts
Synthetic cells in the philosophy of technology

Johannes Achatz

1 Introduction – the value of life

Synthetic biology produces biological artifacts. The artifact might be a synthetic cell with an artificially created and transplanted genome (Gibson et al. 2010). It might be a bacterium "programmed" to produce a kind of bio-fuel (Choi and Lee 2013; Howard et al. 2013). Or it might be a bacterium with a book encoded in its DNA (Church et al. 2012). George Church (self-referentially) encoded his book "Regenesis – How Synthetic Biology Will Reinvent Nature *and* Ourselves" (Church and Regis 2012, pp.269-272) into DNA.

What Church, a leading figure in synthetic biology, promotes in his book is nothing less than a second genesis: a "Regenesis" and "sixth industrial revolution – the information genomics revolution" (Church and Regis 2012, p.210). His colleague Craig Venter sides with him in a similar effort and proclaims, "[w]e have entered now […] the digital age of biology" (Venter 2013, p.1).

The subtitle of Church's book indicates two main approaches to evaluating the biological artifacts of synthetic biology that will herald the new era. National and transnational research and ethic councils generally adopt an *anthropocentric* point of view. They address issues such as: What are the potential risks and chances for humankind? What kinds of regulations are needed to ensure biosafety, biosecurity, and a just distribution of the positive and negative effects of using such technology? While the implementation of safety and security measures is itself uncontroversial, controversies remain about the scope of regulations. For example, Friends of the Earth, representing one hundred and eleven nongovernmental organizations, issued a statement demanding "a moratorium on the release and commercial use of synthetic organisms and their products" (FOE et al. 2012, p.1). The ongoing debate about the use and misuse of products of synthetic biology can be considered a sign of a healthy democratic process of participation in political decision making (M. O'Malley 2013; Knoepffler and M. O'Malley 2014) and will not be our topic here.

The second approach shifts attention from 'ourselves' to 'nature' and can be called *biocentric*. Besides philosophical discussions, this perspective gained considerable legal weight in Switzerland, where the "dignity of living beings" (*Würde der Kreatur*) entered the constitution by means of direct democracy in 1992.[1] The Swiss Federal Ethics Committee on Non-Human Biotechnology, in particular, declared that all living beings, including microorganisms and single bacteria, have a minimal value in themselves: a value so small it is overridden by even the most basic human interests, but nonetheless a value to consider (ECNH 2010, p.18). It is the perspective that matters. Living organisms are not just of value *for* humans, as instruments of human well-being (e.g., intestinal flora); but a world with living organisms – however small they might be – would in general be considered a better world than a world without living organisms. For practical decision making, the Federal Ethics Committee adopts a *hierarchical biocentrism* with humans at the top of the axiological pyramid, above other animals, with animals above plants, and bacteria at the bottom (ECNH 2010, p.18). The resulting recommendations of a hierarchical biocentrism are, insofar as human interests are weighed against the faint values of bacteria, fairly similar to those of anthropocentrism and will also not be our topic here.

This picture might change drastically with advances in protocell research (Rasmussen et al. 2009). As long as existing organisms are used in synthetic biology, their products are well within the reach of laws on *genetically modified organisms* (Schönig 2014, p.36), but once cell-like structures can be produced purely synthetically, they cannot be classified as 'modified organisms' and will need their own category of life-like synthetic cells. So far, this is not the case, and therein might lie part of an explanation of why the artifactual parts of synthetic biology's *biological artifacts* have attracted less ethical and philosophical attention than their biological elements.[2] Keeping in mind George Church's conception of a sixth *industrial revolution* (Church and Regis 2012, p.210) and Craig Venter's proclamation of a

---

[1] "Article 24^{novies} paragraph 3 of the old Federal Constitution, which corresponds to Article 120 of the Federal Constitution revised in 1999" (ECNH 2011, p.6).

[2] The philosophical interest in microbiology is generally quite low; for a notable exception, see (M.A. O'Malley 2014). Philosophical approaches to protocells are few, but they can be subsumed under the broader heading of the philosophy of artificial life (Boden 1996) – the smaller sibling of the philosophy of artificial intelligence. Mark Bedau has done significant ethical work on the topic (Bedau and Parke 2009). Interestingly, the chemical origin of life does not seem to be a topic in the philosophy of chemistry (Weisberg et al. 2011), the discipline in which one would expect a debate on protocells.

digital age of biology (Venter 2013, p.1) the artifactual side does play an important role in the depiction of synthetic biology by some of its leading figures, and seems to be precisely the element that is to bring about the new age.

Thus, neither the dignity of humans nor the "dignity of living beings," but the underrepresented value of artifacts will be our concern and investigated in greater detail. We begin with a take on the definition of artifacts and how artifacts can be evaluated (nonmorally and morally). As will be shown, synthetic biology raises specific issues concerning the standard philosophical approach to defining and evaluating artifacts, namely its strong reliance on the intentionality of an artifact author. To address these issues, a consequentialist approach to artifact definition and evaluation will be put forward. Finally, an answer regarding the possible value of artifacts in synthetic biology is offered, and a typology for different kinds of biological artifacts will be proposed.

2 The value of artifacts

Biological artifacts are the products of synthetic biology. But what exactly *are* artifacts and how can they be evaluated? The standard philosophical approach towards a definition of artifacts and their nonmoral evaluation can be found in Hilpinen's article in the *Stanford Encyclopedia of Philosophy* and reads: "an artifact may be defined as an object that has been intentionally made for some purpose" (Hilpinen 2011). Hilpinen has formulated a set of conditions to allow a precise terminology for the process of artifact identification and evaluation:

- "(A1) An object is an artifact if and only if it has an author."

Following classical approaches to the topic (Aristotle 1987, p.93 - Physics, Book II, 192b9–20), artifacts are defined by how they came into being. Artifacts are not mere natural objects, like sticks and stones, because

- "(DEP) The existence and some of the properties of an artifact depend on an author's intention to make an object of certain kind."

The dependence condition (DEP) ties the artifact to an author's intentionality. An author's productive intention is often documented in drafts, production plans or sketches, and can be matched against the actual character of an artifact. It is important to note that objects that are found, rather than created, might have an instrumental use (e.g., a stick as a hiking pole), but they do not fall under the definition of 'artifacts' according to the term used by Hilpinen (we will return to this exception later on). Animals, however, can also be deemed authors of artifacts, as long as the tools they use are manufactured and not just found, such as in the case of chimpanzees chewing on leaves and mixing them with moss to create sponges (Hobaiter et al. 2014) or crows manufacturing different stick tools (Rutz and St Clair 2012).

In order to distinguish an intended product (e.g., stick tool) from its unintended byproducts (e.g., removed twigs and leaves) Hilpinen adds an evaluative element which leads to a qualitative determination comparing "the intended character of an artifact, its actual character, and a purpose F" (Hilpinen 1995, p.140). Translated into three conditions, an evaluation of artifacts comes to terms with

- "(E1) The degree of fit or agreement between the intended character and the actual character of an object,
- (E2) The degree of fit between the intended character of an object and the purpose F, in other words, the suitability of an object of the intended kind for the purpose F, and
- (E3) The degree of fit between the actual character of an object and the purpose F, that is, the suitability of an artifact for F."

As an example, the intended character of an artifact might be that of a hand tool having the purpose of driving nails into a wall. Its actual character might be that of a stone, a hammer, or a nail gun. If one compares the intended character with the actual character, a nail gun, a hammer, or a stone might be (in a nonmoral sense) better or worse fit for the purpose. A stone is likely easy to obtain and needs little work before it can be used. It might have to be cleaned and broken down into handy pieces. But a stone is unsafe for hammering purposes because it does not offer a firm grip, is hard to handle with precision, and it might simply break. A hammer is harder to manufacture but easier to use. Its force can be modulated while hammering, and it provides a more effective way of handling force. But like the stone, it is still tiresome to use, and the risk of a blue thumb is apparent. A nail gun is most complicated to assemble. Once obtained, it needs electricity and some caution in order to be handled correctly and avoid shooting holes in the wall.

All three artifacts are potentially good implements for sinking nails into a wall, but they each reveal different advantages and disadvantages. To cope with the problem of choosing the (nonmorally) 'best' or 'most suitable' tool, Hilpinen offers two final conditions: a *Success Condition* (SUC) and an *Acceptance Condition* (ACC):

- "(SUC) An object is an artifact made by an author only if it satisfies some sortal description included in the author's productive intention."

- "(ACC) An object is an artifact made by an author only if the author accepts it as satisfying some sortal description included in his productive intention."

The artifact made has to have at least some "degree of fit" with the author's intention, and it has to be accepted as satisfying that intention. Suppose the author's intention was to create a stone tool for hammering that was heavy and big enough but could still be used with both hands. When the author throws a bolder to the ground to break it into a moderately sized piece, the bolder might shatter into pieces too small to be used with both hands. The author now has to decide whether a one-handed stone will suffice for hammering, or whether another attempt should be made to create a suitable tool. The final decision on whether the attempt at creating an artifact is to be considered a success, or at least an acceptable outcome, lies with its author, who must compare his or her initial intentions with the actual character of their productive action in reference to a purpose.

For simple cases of artifact creation where authors evaluate their own products this is a very viable approach. Furthermore, in the context of artworks it has been argued that "in order to be *regarded* (function) as an art work, an object must be regarded as an artifact" (Dipert 1986, p.401) and "[t]o *regard* what is taking place before us as a performance of a Beethoven work, we must regard it as the intentions of the performers to follow what we regard as Beethoven's intentions" (Dipert 1986, p.407). Stretching Hilpinen's definition of artifacts as being *objects* intentionally made for some purpose, even performances might be considered artifacts, although they are not material objects like stones and hammers, but intended *events* represented by a musical score. Houkes and Vermaas (2010, p.2) have therefore classified Hilpinen's approach as one amongst the "*intentional function theories*". Similar suggestions, focusing on the notion of *function*, have been made concerning biological artifacts (Krohs and Kroes 2009).

Three potentially good implements that can serve as hand tools for the purpose of driving nails into a wall have been identified. We know *who* will decide and evaluate the artifacts: the author. However, the question of *how* it will be decided which artifact should be used has so far been outlined but not answered sufficiently with the insistence on taking the author's intentions into account:

Which artifact is chosen depends also on personal preference (e.g., if someone is left-handed or right-handed), personal experience (e.g., if they have never used a nail gun), and beliefs regarding environmental factors (e.g., concrete wall or plywood wall). These aspects are not addressed directly in Hilpinen's account, but they might be assumed to be already contained in the author's intention, which leaves us with a conundrum:

- If personal preference, experience and beliefs are *not* included in the notion of intentionality: intentions are not sufficient to evaluate artifacts.
- If personal preference, experience and beliefs *are* included in the notion of intentionality: only the author can decide if his or her product is suitable to be considered an artifact *at all*, let alone if it can be reliably evaluated.

The undefined, broad notion of intentionality becomes problematic especially when one attempts to move away from a nonmoral evaluation towards a moral evaluation of artifacts.

3   The problematic first-person perspective on biological artifacts

In practical philosophy and ethics, in particular, the first-person perspective on artifact creation forces a person B who is affected by another person A's creation of an artifact to consider A's *intentions* before the result of a productive action can be identified as an artifact and subjected to (moral) evaluation. Several problems in the evaluation of person A's artifact creation arise from that conception, especially if the author and judge of an artifact are two different persons:

1. Person A's creative intention might not be available to person B. Archaeologists and anthropologists often have to guess whether excavated stones and bones had specific functions. Animals who used tools can also not be questioned about their intentions – these have to be surmised and verified in experiments. Or to use another example: an agent, person A, might simply be unwilling to communicate any intentions to person B. Person A can even tell downright lies. Following Hilpinen's reasoning, person A could – by withholding communication of their 'private' intentionality – effectively block B's moral evaluation: a knife crafted by person A might fail to be of good use for *cutting* onions but, contrary to the intentions expressed by person A, could still be a great instrument for *stabbing* person B.
2. Entities might be declared objects of art (a posteriori) rather than be (physically) created as such; their coming-to-be as art does not have to be the direct result of the creator's artistic intentionality. For example, John Cage's musical compositions often rest on probability or rely on the *I Ching*. The authorship of the actual musical experience becomes debatable. The (empty) sheet of the musical

score of *4'33"* famously escapes the common belief of what makes a musical piece or an artistic performance of a composer's work.
3. The intended character, actual character, and also the purpose of the artifact's creation might be unspecific. Especially in new biotechnological research as conducted in the field of synthetic biology (Annaluru et al. 2014; Gibson et al. 2010), a methodology characterized as "kludging" (M.A. O'Malley 2011)[3] makes it difficult to apply Hilpinen's criteria to evaluating the creation of artifacts or "biofacts" (Karafyllis 2003). iGEM.org, for example, stages a competition for post-genomic biofact design. Hobby biologists have formed groups (DIYbio.org). And artists are exploring new possibilities with BioArt (notinventedbynature.org) (Davis 1996; Kerbe and Schmidt 2013). All three communities use forms of kludging in genetic and biological experimentation that methodologically exclude their own intentions in the creation of artifacts. Creation comes first, and they subsequently attempt to recognize and evaluate artifacts and their possible characters and functions. Such playful approach points to artifacts being yet another example of *family resemblance* (Wittgenstein 1986, p.32, par.66-67) with different artifacts showing "similarities overlapping", but not all artifacts sharing a common trait.
4. The sharp distinction of artifacts from non-manufactured tools as well as from unintended byproducts (e.g., waste, scrap, pollution) is debatable. First, the unintended byproducts of the extensive use of technology caused the ecological crisis in the nineteen-sixties (Meadows et al. 1972) and led to a rethinking of the status of nature, humans, and technology, and to the birth of environmental ethics in the nineteen-seventies (Jonas 1979; Næss 1973; Passmore 1974; Singer 1975; Stone 1974) as well as to political movements that formed into green parties all around the globe. Second, it is often possible to find a use for unintended byproducts of artifact creation. For example, sawdust or wood shavings can be turned into sought-after isolation and packing material. Considering the possible consequences of artifact creation, such as the generation of waste and byproducts, is a central part of responsible engineering and a standard maneuver in the planning stage preceding artifact creation.
5. The exclusion of non-manufactured tools like sticks used as hiking poles from the definition of an artifact while including artworks, instruments, machines, computers, and even buildings or entire industrial product assembly lines is debatable.

---

3   But still different from "tinkering" (Morange 2013, p.322).

Hilpinen's focus on creator intentionality to define artifacts is thus problematic in cases where the author of an artifact is not the person evaluating it. Industrial production plans might show the intended character and purpose of an artifact. However, the examples of archeology, art, and synthetic biology demonstrate the problematic nature of depending heavily upon preceding intentionality for artifact evaluation. Such a focus is especially incapable of adequately guiding necessary *moral* judgments on artifacts created in new biotechnological sciences.

4   Other intentional function theories

Lately, several solutions to these kinds of problems have been developed, namely: the *ICE theory* of Houkes and Vermaas (2010), which draws on the intentional, casual-role, and evolutionist function theories. The ICE-function theory focuses on *technical* functions, but promises to be expanded to the domains of art and biology. Similarly, Peter Kroes omits biotechnical artifacts from his considerations but includes technical artifacts (and also, by way of example, artworks) in what he calls a *dual nature* conception of them as "physical constructions with a technical (practical) *function*" (Kroes 2012, p.4). Both of these approaches still rely heavily on intentionality but go beyond Hilpinen's standard approach by including physical and action-theoretical elements to better grasp societal effects and avoid the common shortcomings of a purely *intentional function theory* of artifacts.

A theory of artifact creation being an "action type" has been proposed by Gregory Currie (1989) and criticized by Hilpinen (1992, p.73). The proposed "action type" is a historical concept: "the appreciation of art works is the appreciation of a certain kind of achievement. Now it is relevant, as I have said, in finding out what someone's achievement is, to know what others have done" (Currie 1989, p.72). In short, Currie's approach is a sophisticated way of asking: who did it first, and under what circumstances, when trying to evaluate a piece of art;[4] an approach that is also not widely applicable to products of synthetic biology because its spectrum of considered artifacts is too narrow and focuses on works of art exclusively.

---

4   Gregory Currie's approach is quite reminiscent of Walter Benjamin's theory from 1935/1939. Artworks derive their authenticity from having a unique appearance in the here and now. Benjamin criticizes technical reproduction (print, photo, film) for eroding this tradition and destroying the originality and authenticity of art (Benjamin 1991). Benjamin, however, is not favored in Currie's work.

5 A third-person perspective on artifact creation

Most of the above-mentioned problems stem from concentrating on elements preceding artifact creation. Intentionality is used as a black box containing personal preferences, personal experience, and beliefs considered necessary for artifact evaluation. As this black box is situated in the author of an artifact, the sole sovereignty on artifact evaluation lies with the author.

According to John Dewey, consequences as opposed to intentions should move to the center of philosophical attention when valuating artifact-creating actions (Dewey 1922, p.352). This suggestion is at the outlines of a vigorous philosophical debate on the nature of truth and the status of epistemology as a primary field of research in philosophy. Neopragmatists like Richard Rorty, echoing Dewey, have raised heavy philosophical artillery when criticizing the focus on intentionality and strong emphasis on an author's *mind* because they believe it represents a philosophical tradition of favoring the theoretical over the practical, the mind over bodily action, the absolute over processes of growth and decay, or in Kantian terms: the categorical over the hypothetical (Rorty 1979).[5]

Claims of this magnitude touch on our topic, but transcend the issue of artifact creation we set out to examine. What we can take from Dewey, and what will currently help us in our project, is pinpointing artifact creation on human action instead of aiming at a fixed definition of artifacts as materializations of an author's ulterior intentions. I have argued extensively (Achatz 2013, pp.159-199) for a wider definition of technology that is not just a name for a class of objects resulting from human activity but states that technology is most of all a stepping stone for further action and activity. Technology, in short, is externalized human action.

Artifacts can be seen as enabling an action (drilling a hole) if they can be considered to fulfill what Searle calls *agentive functions* (Searle 2011, p.58). *Attributions* of agentive functions to entities are *subsequent* attributions. Intentionality still plays an important role, but it is not necessarily tied to the author of an artifact. Instead, a wider definition of functions as offered by Searle (Searle 1995, pp.16ff.) is less author relative, but rather *observer* relative:

---

5   The philosophical movement called pragmatism drives attacks against what its protagonists see as epistemology's exaggerated claims of truth (Rorty 1979) and against the belief in a sharp distinction between facts and values (Putnam 2003). Instead, pragmatists maintain that "valuing" is part of the standard setup of human activity, and while striving for (neutral) facts is highly desirable, the outcomes are "best guesses" and open to change through new findings and progress in scientific research (Dewey 1998, p.19).

- Agentive functions: "the ascription of the function ascribes *the use to which we intentionally put* these objects" like "This stone is a paperweight."
- Non-agentive functions: "naturally occurring causal processes to which we have assigned a purpose, e.g., the function of the heart is to pump blood."

Agentive and non-agentive functions form a schema to be used to ascribe and predict the behavior of an entity in a system:

- "1. Whenever the function of X is to Y, X and Y are parts of a *system* where the system is in part defined by *purposes, goals, and values generally*. This is why there are functions of policemen and professors but no function of humans as such – unless we think of humans as part of some larger system where their function is, e.g., to serve God.
- 2. Whenever the function of X is to Y, then X is *supposed* to cause or otherwise result in Y. This normative component in functions cannot be reduced to causation alone, to what in fact happens as a result of X, because X can have the function of Y-ing even in cases where X fails to bring about Y all or even most of the time. Thus the function of safety valves is to prevent explosions, and this is true even for valves that are so badly made that they in fact fail to prevent explosions, i.e., they *malfunction*." (Searle 1995, p.19)

Searle's definition of function matches well with the definition of technical action as being the intended use of certain means to realize ends according to values (Hubig 1993, p.43). The human mind does the 'intending', and thus sticks and stones can have 'functions'; but it does not have to be the mind of the author of an artifact. In short, a consequentialist approach allows for moral evaluation via the attribution of responsibility: The consequences of actions are attributed to an author in reference to certain values (Kunzmann 2010, p.2). A person is responsible for the foreseeable consequences of his or her actions (Weber 1946, p.121). The focus shifts from the (first-person) 'intended function' to its (third-person) possible uses. While an author him/herself can conduct a (private)[6] moral evaluation of their first-person actions, an audience or bystander can only rely on a (consequentialist) evaluation

---

6   Max Weber's term *Gesinnungsethik* expresses precisely what is meant by "private moral evaluation", but it has been poorly translated as "absolute value ethics" (Weber 1949, p.21) or "ethic of ultimate ends" (Weber 1946, p.120). While the German term encompasses an ethic of conviction and conscience, an ethic of disposition, and direction of the will, or duty, it is not exclusively an ethic of absolute values.

of the possible uses of an artifact (if it is clearly supposed to be of technical use) or possible interpretations (if the artifact is clearly an object of art).

6 A consequentialist approach to a moral evaluation of artifacts

On revisiting the above critique (1-5) of Hilpinen's standard approach to artifact creation and evaluation and applying the consequentialist perspective, the following picture emerges:

1. Possible uses of an artifact are obtained by the ascription of agentive functions. Availability of an author's intention can be of great help in evaluating an artifact, but the perspective of a *user* is sufficient. Foreseeable consequences of artifact use are the hallmark of responsibility in designing and using artifacts.
2. The declaration that something is an object of art, rather than the creation of an object of art, marks the very limit of what a definition of an artifact is supposed to contain. Again, the foreseeable consequences of an artifact's use and *interpretation* indicate the responsibility relationships involved in artifact creation. John Cage's silent piece *4'33"* does not pose a challenge to a moral evaluation via responsibility. Philosophically, his composition might be better defined as a kind of social artifact, which can only be considered an artifact because it has been declared to be one. The silence of 4'33" is presented as an artwork and only therefore can it be recognized as such. Every time the piece is performed, the silence (and noise made by the audience, etc.) becomes 'music' as indicated in John Cage's musical score. It can be called, to cite Searle once again, a "standing Status Function Declaration" (Searle 2011, p.97).
3. Responsibility as the attribution of consequences of actions to an author in reference to certain values (Kunzmann 2010, p.2) does stretch to include playful action and "kludging" (M.A. O'Malley 2011), as practiced in biological artifact creation by iGEM participants, DIY biologists and BioArtists. To be considered an artifact, it is sufficient for a created or utilized entity to be seen to exercise agentive functions. A bacterium in which the text of a book is stored is a prime example of the artifactual use of an entity by ascription of an agentive function (this particular organism stores this particular book) and externalized action (external memory). A more detailed differentiation of types of artifacts can be obtained by sorting the externalized means and purposes of an artifact, as will be shown in the next section of this paper.

4. Non-manufactured tools can be considered artifactual if they can be used to execute agentive functions. Byproducts of artifact creation are likewise artifacts. They bear the mark of human ingenuity just as much as the intended final product and as "foreseeable consequences" (Weber 1946, p.121) of artifact creation, the author of the artifact is responsible for them. Identifying foreseeable consequences can become a task of severe complexity, given that technological artifacts are, as industrial products, often the outcome of a long chain of supply, of manufacturing, distribution, division of labor, shipping, and diverse use. A field of (social) science is devoted to identifying the consequences of new technologies on a societal scale, namely Technology Assessment.
5. As mentioned in (3), a typology of artifacts can be developed utilizing function ascriptions and further differentiating levels of complexities in artifacts, represented by means and purposes 'built' into an artifact as sortal criteria.

7   Sorting kinds of artifacts and biofacts with agentive functions

The attribution of agentive functions to entities can be paralleled with the attribution of responsibility. Differentiating the range of *agentive functions* that can be attributed to an entity allows a classification of different kinds of technical entities. Simple agentive functions can be executed without manufacturing (e.g., a stick used as a hiking pole). More complex agentive functions can be realized by externalizing *means*, *purpose* or a *course of action* to a larger extent into an entity. In contrast to Hilpinen's use of 'artifact' as a generic term for all objects intentionally made for a purpose, 'artifacts' are treated as just one simple kind of technical entity, followed by tools (e.g., hammer), machines (e.g., washing machine), and artifactual information systems (e.g., protocells).

- Artifacts – The minimum criterion for an entity to be considered an artifact is that it has been created or is used to execute agentive functions. Entities like art or waste products do not necessarily have to be created intentionally or for a specific purpose to be considered artifacts. A proof-of-concept is a good example of an artifact that is not useful in itself but, when interpreted as (or declared to be) a technological milestone, can nonetheless be of immense value and have far-reaching consequences (e.g., to acquire more research funding).
- Tools – A tool has a specific purpose but still comes very close to unenhanced human bodily action. A hammer can be used to sink nails into a wall, and the

action of hammering can be altered, slowed down, sped up, or stopped at any time. Hence simple tools have been considered extensions of the human body (Kapp 1877), allowing a similar kind of control for tools as for executing tasks with bare hands and feet.
- Machines – While a tool is already built for a specific need or task, machines exhibit a higher level of complexity. They do not represent a single means or purpose, but embody a *course of action* externalized in a (fixed) process. The means have already been pre-picked by the author/creator of the machine, and the user of a machine can only choose amongst, for example, different washing programs, but not how the machine heats the water. Unlike simple tools, the workings of a machine are not always obvious and cannot be as easily altered or stopped.
- Artificial information systems – Representing another level of complexity in technological entities, artificial information systems are systems able to 'choose' amongst several built-in means for a purpose. For instance, a 'web-crawler bot' or a computer virus has a range of available courses of action at its disposal when it encounters specific situations. Depending on the circumstances, different courses of action can be applied by the system (or 'chosen' by the 'agent-program', as the creators of artificial intelligence are fond of saying). Artificial information systems can – and this makes them similar to living organisms in a certain respect – carry within them a description of themselves or their purpose, because the program code of a computer virus can be used to multiply the virus, similar to the way in which the genetic code of an organism is passed on in reproduction. The drawing of a parallel between program code and genetic information is a classic theme in the theory of biology (Schrödinger 2004, pp.111f.) and an important, or *the* most important factor in the description of organisms as computing machines. As Sydney Brenner, a South African biologist and Nobel Laureate writes: "Arguably the best examples of Turing's and von Neumann's machines are to be found in biology. Nowhere else are there such complicated systems, in which every organism contains an internal description of itself. The concept of the gene as a symbolic representation of the organism — a code script — is a fundamental feature of the living world and must form the kernel of biological theory" (Brenner 2012, p.461).
- Nature-identical systems – Our list of technical entities of growing complexity is completed by artificial systems 'grown' by human ingenuity and imitating nature perfectly. Such systems do not yet exist, but protocells might one day become chemically identical to their biological role models, just as one finds nature-identical flavoring substances in yogurt or artificially created insulin. Such artificial information systems could become the definite bridge between

artificial and biological systems, being discernible as 'biological artifacts' not by their outlook, but just by their genesis. If protocells are ever developed to such an extent, they will raise serious questions, especially for *biocentric* ethical positions that attribute a moral value to every living organism (ECNH 2010, p.18) and rely on the difference between natural organisms and artifactual systems.

Nicole Karafyllis has developed, similar to the above list of technical entities, a typology of biotic artifacts she calls "biofacts" (Karafyllis 2003; 2006). Using (natural) growth and reproduction as the sortal category, she differentiates between different levels of phenotypic and genotypic intervention in the growth of an organism. By complementing the growth and reproduction of organisms with our categories of technical entities, a typology of different kinds of biofactual uses of organisms emerges:

- Natural life form – The ascription of non-agentive functions allows the description of natural life forms in terms of means and purposes (e.g., plants growing towards the sun for more effective photosynthesis).
- Unaltered biofacts – Organisms can have agentive functions as well as non-agentive functions if they serve a human purpose. Ladybugs keeping down aphids on home-grown tomatoes are welcomed by gardeners. Ladybugs can be bred and purchased specifically for that purpose, but in many regions they also occur naturally, thus being of instrumental use to humans, while their growth and replications remains untouched.
- Reshaped biofacts – The growth of an organism can be altered for human purposes in a *reversible* way if the intervention remains on the level of phenotypic reshaping, like the artful pruning of a bonsai tree. Although the bonsai's growth is evidently restrained for a human purpose, the tree as well as its progeny will continue to grow once human intervention ceases.
- Genetically reproduced biofacts – Cloning marks a more far-reaching intervention in the growth and reproduction of an organism. While it does not change the organism itself, cloning is a *long-term* intervention, or externalization of human means, that prevents 'normally' occurring change in the replication of an organism for human purposes.
- Genetically modified biofacts – Finally, the growth and replication of an organism can be *irreversibly* adjusted to human purposes. Extensive breeding and selection can lead to this effect, but genetically modified organisms created by different methods of gene technology and Craig Venter's "synthetic cell" (Gibson et al. 2010) from the specific field of synthetic biology are probably the best known examples of the artifactualization of organisms, reaching a new depth of exter-

nalization of human means and ends and causing irreversible consequences for the growth and replication of an organism. The playful experiment of a book encoded in the DNA of an organism (Church et al. 2012) transforms the very non-agentive means of a genome to execute an agentive function for human purpose. The bacterium becomes a complex biofact, an organism used as an artificial information system.

The "programmed" synthetic cells of synthetic biology (Gibson et al. 2010) and the artificially created protocells (Rasmussen et al. 2009) will eventually merge. The convergence of biological and informational sciences, as pictured by George Church's sixth industrial revolution (Church and Regis 2012, p.210) and Craig Venter's digital age of biology (Venter 2013, p.1) reflects a longing for a biofactual future which will, in time, make further philosophical distinctions necessary.

In particular, the notion of 'programming' cells, and Craig Venter's insistence on having given "the proof, through construction of a synthetic chromosome and hence a synthetic cell, that DNA is the software of life" (Venter 2013, p.7) bestows computing metaphors and artifactual vocabulary with a growing importance in biotechnologies. Questions concerning what the ontological 'nature' of a computer program is, what kind of 'hardware' the 'wetware' of a cell can be, and the technical use of biofacts in industrial production will demand more attention not only from applied bioethics, but also from the philosophy of technology in general.

Returning to the question of the value of artifacts and picking up the thought experiment from the introduction, a lifeless world *with* artifacts compared to a lifeless world *without* artifacts, can hardly be seen as being of 'value'. A priceless piece of art or a clever machine can be of (considerable) *instrumental* and *societal* value, but without subjects who are capable of appreciating them the artifacts themselves strive for nothing, break down, decay, and can hardly be deemed valuable. This result of our investigation might not surprise, but it is nevertheless of particular importance in distinguishing moral and non-moral values attributed to biofacts.

8 Conclusion

Philosophical debates on nature and technology span millennia. Nature – and the emergence of living organisms – remains open to interpretation and investigation. Just as the proponents of natural ethics assume and defend a value in living organisms, so too does the description of cells as biological computers, as in "DNA software builds its own hardware" (Gibson et al. 2010), bridge the interpretational gap (Achatz 2014). Both approaches are not just embellishments, but metaphysical positions reaching beyond "good reductionist science" (Venter 2013, p.109) and offering a framework for interpreting the advances in synthetic biology. Sorting the instrumental values of artifacts and clarifying the responsibility relationships involved in the creation of biological artifacts is a necessary and demanding societal task, especially in light of the fact that synthetic biology is still currently in the proof-of-principle stadium and the promises of a "sixth industrial revolution" (Church and Regis 2012, p.210) and a "digital age of biology" (Venter 2013, p.1) remain as yet unfulfilled.

In the late 1970s Hans Jonas described how the consequences of human technological civilization have reached both the deepest levels of the sea (pollution) and the highest regions of the planet (holes in the ozone layer, scrap of defunct satellites in space). Because, presumably, no place has been left untouched by the far-reaching consequences of human technological activity, Hans Jonas mused that the world had become a "total artifact" (Jonas 1984, p.33), and as such an artifact had fallen to human responsibility. Philosophical interest in artifacts has not become a central topic of practical philosophy; however, it is time for a change.[7]

References

Achatz, J. (2013). *Synthetische Biologie und ‚natürliche' Moral: ein Beschreibungs- und Bewertungszugang zu den Erzeugnissen Synthetischer Biologie*. Freiburg: Alber.

Achatz, J. (2014). Framing ‚Nature' - Synthetische Biologie schreibt (ihre) Geschichte. In: J. Achatz, N. Knoepffler (eds.), *Lebensformen – Leben formen. Ethik und Synthetische Biologie* (pp.83-100). Würzburg: Königshausen & Neumann.

Annaluru, N., Muller, H., Mitchell, L.A., Ramalingam, S., Stracquadanio, G., Richardson, S.M., ... Chandrasegaran, S. (2014). Total Synthesis of a Functional Designer Eukaryotic Chromosome. *Science, 344*, 55-58. doi: 10.1126/science.1249252.

Aristotle (1987). Physics. In: J.L. Ackrill (ed.), *A new Aristotle reader* (pp.81-131). Oxford: Clarendon.

Bedau, M., & Parke, E.C. (eds.) (2009). *The ethics of protocells – moral and social implications of creating life in the laboratory*. Cambridge: MIT Press.

---

[7] Acknowledgments: Part of the research presented in this paper was conducted in the project *Framing 'Nature': The moral standing of life forms and life-artifacts* led by Prof. Dr. mult. Nikolaus Knoepffler and funded by the German *Federal Ministry of Education and Research*. I am also grateful to my colleague Martin O'Malley (PhD) for joint work in said project, and to both him and Christiane Burmeister for commenting on a draft of this text.

Benjamin, W. (1991). Das Kunstwerk im Zeitalter seiner technischen Reproduzierbarkeit. In: R. Tiedemann, H. Schweppenhäuser (eds.), *Gesammelte Schriften Bd. 1 Abhandlungen Teil 2* (Vol. I, pp.471-508). Frankfurt/Main: Suhrkamp.

Boden, M.A. (ed.) (1996). *The philosophy of artificial life.* Oxford: Oxford University Press.

Brenner, S. (2012). Turing centenary: Life's code script. *Nature, 482*(7386), 461-461.

Choi, Y.J., & Lee, S.Y. (2013). Microbial production of short-chain alkanes. *Nature, 502*(7472), 571-574. doi: 10.1038/nature12536.

Church, G.M., & Regis, E. (2012). *Regenesis : how synthetic biology will reinvent nature and ourselves.* New York: Basic Books.

Church, G.M., Gao, Y., & Kosuri, S. (2012). Next-Generation Digital Information Storage in DNA. *Science, 337*(6102), 1628. doi: 10.1126/science.1226355.

Currie, G. (1989). *An Ontology of Art* (Vol. 40). London: St. Martin's Press.

Davis, J. (1996). Microvenus. *Art Journal, 55*(1), 70-74. doi: 10.2307/777811.

Dewey, J. (1922). Valuation and Experimental Knowledge. *The Philosophical Review, 31*(4), 325-351. doi: 10.2307/2179099.

Dewey, J. (1998). From Absolutism to Experimentalism. In: L. A. Hickman (ed.), *The essential Dewey - Pragmatism, education, democracy* (Vol. 1, pp.14-21). Bloomington, Ind.: Indiana University Press.

Dipert, R.R. (1986). Art, Artifacts, and Regarded Intentions. *American Philosophical Quarterly, 23*(4), 401-408. doi: 10.2307/20014165.

ECNH (2010). Federal Ethics Committee on Non-Human Biotechnology. *Synthetic biology - Ethical considerations.* Bern.

ECNH (2011). Federal Ethics Committee on Non-Human Biotechnology. *Report of activities 2008 - 2011.* Bern.

FOE, CTA, & ETC. (2012). Friends of the Earth, International Center for Technology Assessment, & ETC-Group. *The Principles for the Oversight of Synthetic Biology.* http://www.synbioproject.org/process/assets/files/6620/_draft/principles_for_the_oversight_of_synthetic_biology.pdf. Accessed: 19 May 2015.

Gibson, D.G., Glass, J.I., Lartigue, C., Noskov, V.N., Chuang, R.Y., Algire, M.A., ... Venter, J.C. (2010). Creation of a bacterial cell controlled by a chemically synthesized genome. *Science, 329*(5987), 52-56. doi: 10.1126/science.1190719.

Hilpinen, R. (1992). On artifacts and works of art. *Theoria, 58*(1), 58-82.

Hilpinen, R. (1995). Belief Systems as Artifacts. *The Monist, 78*(2), 136-155.

Hilpinen, R. (2011). Artifact. In: E.N. Zalta (ed.), *The Stanford Encyclopedia of Philosophy* (Winter 2011 ed.).

Hobaiter, C., Poisot, T., Zuberbühler, K., Hoppitt, W., & Gruber, T. (2014). Social Network Analysis Shows Direct Evidence for Social Transmission of Tool Use in Wild Chimpanzees. *PLOS Biology, 12*(9), e1001960. doi: 10.1371/journal.pbio.1001960.

Houkes, W., & Vermaas, P.E. (2010). *Technical Functions: On the Use and Design of Artefacts.* Series: Philosophy of Engineering and Technology 1. Dordrecht: Springer Netherlands. doi: 10.1007/978-90-481-3900-2.

Howard, T.P., Middelhaufe, S., Moore, K., Edner, C., Kolak, D.M., Taylor, G.N., ... Love, J. (2013). Synthesis of customized petroleum-replica fuel molecules by targeted modification of free fatty acid pools in Escherichia coli. *Proceedings of the National Academy of Sciences*, 1-6. doi: 10.1073/pnas.1215966110.

Hubig, C. (1993). *Technik- und Wissenschaftsethik : ein Leitfaden.* Berlin: Springer.

Jonas, H. (1979). *Das Prinzip Verantwortung : Versuch einer Ethik für die technologische Zivilisation.* Frankfurt/Main: Insel.
Jonas, H. (1984). *Das Prinzip Verantwortung : Versuch einer Ethik für die technologische Zivilisation.* Frankfurt/Main: Suhrkamp.
Kapp, E. (1877). *Grundlinien einer Philosophie der Technik: zur Entstehungsgeschichte der Cultur aus neuen Gesichtspunkten.* Braunschweig: Westermann.
Karafyllis, N.C. (2003). *Biofakte : Versuch über den Menschen zwischen Artefakt und Lebewesen.* Paderborn: Mentis.
Karafyllis, N.C. (2006). Biofakte - Grundlagen, Probleme, Perspektiven. *Erwägen Wissen Ethik EWE, 17*(4), 547-558.
Kerbe, W., & Schmidt, M. (2013). Splicing boundaries: The experiences of bioart exhibition visitors. *Leonardo, 48*(2), 128-136. doi: 10.1162/LEON_a_00701.
Knoepffler, N., & O'Malley, M. (2014). Synthetische Biologie - Ethische Überlegungen. In: J. Achatz & N. Knoepffler (eds.), *Lebensformen – Leben formen. Ethik und Synthetische Biologie* (pp. 55-69). Würzburg: Königshausen & Neumann.
Kroes, P. (2012). *Technical Artefacts: Creations of Mind and Matter* (Vol. 6). Dordrecht: Springer Netherlands.
Krohs, U., & Kroes, P. (2009). *Functions in biological and artificial worlds: comparative philosophical perspectives.* Cambridge: MIT Press.
Kunzmann, P. (2010). Die Verantwortung des Verbrauchers und einige ihrer Grenzen. *Journal für Verbraucherschutz und Lebensmittelsicherheit, 5*(1), 1-5. doi: 10.1007/s00003-009-0528-8.
Meadows, D.H., Meadows, D.L., Randers, J., & Behrens III, W.W. (1972). *The limits to growth : a report for the Club of Rome's project on the predicament of mankind.* New York: Universe Books.
Morange, M. (2013). Comparison Between the Work of Synthetic Biologists and the Action of Evolution: Engineering Versus Tinkering. *Biological Theory, 8*(4), 318-323. doi: 10.1007/s13752-013-0134-y.
Næss, A. (1973). The shallow and the deep, long-range ecology movement : a summary. *Inquiry, 16*, 95-100.
O'Malley, M. (2013). Value Ethics: A Meta-Ethical Framework for Emerging Sciences in Pluralistic Contexts. In: C. Baumbach-Knopf, J. Achatz, N. Knoepffler (eds.), *Facetten der Ethik* (pp.73-93). Würzburg: Königshausen & Neumann.
O'Malley, M.A. (2011). Exploration, iterativity and kludging in synthetic biology. *Comptes Rendus Chimie, 14*(4), 406-412. doi: 10.1016/j.crci.2010.06.021.
O'Malley, M.A. (2014). *Philosophy of microbiology.* Cambridge: Cambridge University Press.
Passmore, J.A. (1974). *Man's responsibility for nature: ecological problems and Western traditions.* New York: Scribner.
Putnam, H. (2003). *The collapse of the fact/value dichotomy and other essays* (2nd print ed.). Cambridge: Harvard University Press.
Rasmussen, S., Bedau, M.A., Chen, L., Deamer, D., Krakauer, D.C., Packard, N.H., & Stadler, P.F. (2009). *Protocells : bridging nonliving and living matter.* Cambridge: MIT Press.
Rorty, R. (1979). *Philosophy and the mirror of nature.* Princeton, N.J.: Princeton University Press.
Rutz, C., & St Clair, J.J.H. (2012). The evolutionary origins and ecological context of tool use in New Caledonian crows. *Behavioural Processes, 89*(2), 153-165. doi: dx.doi.org/10.1016/j.beproc.2011.11.005.

Schönig, B. (2014). Synthetische Biologie: Biologische Sicherheit und Risikobewertung durch die ZKBS. In: J. Achatz, N. Knoepffler (eds.), *Lebensformen - Leben formen ; Ethik und Synthetische Biologie* (pp.26-36). Würzburg: Königshausen & Neumann.

Schrödinger, E. (2004). *Was ist Leben? Die lebende Zelle mit den Augen des Physikers betrachtet*. München: Piper.

Searle, J.R. (1995). *The construction of social reality*. New York: Free Press.

Searle, J.R. (2011). *Making the social world: the structure of human civilization*. Oxford: Oxford University Press.

Singer, P. (1975). *Animal liberation: a new ethics for our treatment of animals*. New York: New York Review.

Stone, C.D. (1974). *Should trees have standing? Toward legal rights for natural objects*. Los Altos, Calif.: W. Kaufmann.

Venter, J.C. (2013). *Life at the speed of light: from the double helix to the dawn of digital life*. New York: Viking.

Weber, M. (1946). Politics as a Vocation. In: H.H. Gerth, C.W. Mills (eds.), *From Max Weber: Essays in sociology* (pp.77-128). New York: Oxford University Press.

Weber, M. (1949). The Meaning of "Ethical Neutrality" in Sociology and Economics. In: E. Shils, H. Finch (eds.), *Max Weber on the methodology of the social sciences* (1st ed., pp.1-47). Glencoe, Ill.: Free Press.

Weisberg, M., Needham, P., & Hendry, R. (2011). Philosophy of Chemistry. In: E.N. Zalta (ed.), *The Stanford Encyclopedia of Philosophy* (Winter 2011 ed.).

Wittgenstein, L. (1986). *Philosophical investigations* (trans. by G.E.M. Anscombe, 3rd ed.). Oxford: Blackwell.

II
Public Good and Private Ownership.
Social and Legal Ramifications

Legal Aspects of Synthetic Biology

Jürgen Robienski, Jürgen Simon and Rainer Paslack

1 Introduction

Synthetic biology is designated as a new and most promising technology. Novel technologies always raise the question of whether new laws are necessary. This question has to be answered in the affirmative if the new technology brings about completely new potential risks that would urge the legislative authority to issue a special law for it (Dederer 2010, pp.71f.).

Since 2006/2007, great efforts have been made worldwide to find a consistent definition of the scientific area of synthetic biology (Boldt et al. 2009, p.8) and to reach consensus on what constitutes the core idea and the novelty of synthetic biology (Catenhusen 2011, p.85; Sauter 2011, p.23). Nevertheless, there is still no consistent definition of synthetic biology in the natural sciences as yet.

In the German technical literature, the term synthetic biology appeared from the 1980s onwards in connection with the debate on the regulation of genetic engineering. Synthetic biology was mostly seen as being synonymous with modern genetic engineering (Herbig 1978, p.10; Lengeler 1988, p.17; Herdegen 2012; Hohlfeld 1986, pp.550-560; Hohlfeld 1984, pp.550-560; Hohlfeld 1990). As early as 1978 Herbig wrote: "Biology has reached the critical stage of a science: They constructed nature. The era of synthetic biology has begun. Engineering and biology combine to genetic engineering, [...] Biologists become creators, more precisely, to designers of new creatures." In these early texts the same chances, goals, promises and open questions were already discussed under the concept of synthetic biology as a synonym for genetic engineering.[1] Herbig even illustrated the production of a

---

1 Hohlfeld, R. (1988, p.61): "In fact today physicists, chemists and molecular biologists using gene technology, chemical and biochemical synthesis methods and highly advanced automation techniques can construct biological agents, genes, cell-like membrane

synthetic gene which does not occur in nature using the example of the synthetic somatostatin gene made by Itakura and Riggs (Herbig 1978, pp.191f.). In light of the above, it is not surprising that there is broad agreement that 95-98 % of what is declared 'synthetic biology' is a straightforward continuation of modern molecular biology, genetic research or genetic engineering, and, for that reason, there is nothing new (Müller-Röber 2011, p.70) about it.

This assumption is also supported by the fact that the scope and legal definitions contained in the German Genetic Engineering Act (GenTG), particularly with regard to GMOs, covers most of the subgroups of synthetic biology named by Schmidt. Schmidt cites the statement that synthetic biology is "the design and construction of new biological components, instruments and systems and the re-design of already existing natural biological systems for beneficial aims" as the most common "definition" in the scientific community (Schmidt 2011, pp.111f.; see also http://syntheticbiology.org/Who_we_are.html).

Based on this definition, Schmidt defines five subgroups as the most important areas of application of synthetic biology (Schmidt 2011, pp.112f.):

a. DNA synthesis: chemical construction of genetic codes based on the matrix of a genetic code of an existing organism (with known nucleic acids);
b. DNA-based biological circuits: transfer of complete biological systems made of biobricks;
c. Minimal genome or minimal life form (top-down process);
d. Protocells: living cells which are constructed new from the bottom up;
e. Xenobiology: creation of orthogonal biological systems that do not occur in nature, based on biochemical principles that do not occur in nature (XNA).

These five subgroups can be reduced to three main elements: modification, copying, and new creation of 'life'.

The only regulatory loophole that could be identified concerns the element of 'new creation'; the question is therefore whether a 'de novo' synthetically created cell or orthogonal biological system which does not occur in nature is also a "biological unit which is capable to propagate or to exchange genetic material" pursuant to the GenTG (Engelhardt 2010, p.21). The Central Commission for Biological Safety (ZKBS) in Germany noted in its recent interim report from November 06, 2012 that most scientific approaches in synthetic biology fall within the scope of the

---

vesicles and organisms with a completely new genetic map, which nature has never seen before."; Eberbach (2012, p.13 (12)) referring to the historical debate on the risks of genetic engineering.

GenTG. Only novel living systems such as artificial cells (bottom-up approach) without precedent in nature are not covered by the GenTG (ZKBS 2012, p.8). In this respect, a small clarifying supplement to the legal definition of the term organism in the GenTG would suffice to close the loophole. The supplement could be phrased as follows: "any biological unit which is capable of propagating or transferring genetic material, including microorganisms, and *any biological units created by technical means which do not occur under natural conditions and which contain genetic material that does not occur naturally.*" A supplement to that effect would clarify that synthetically produced or modified organisms or biological units, and even the use of naked, synthetically produced DNA would definitely fall within the scope and field of control of the GenTG.[2]

Although synthetic biology does not seem to be a fundamentally new technology – in particular in a legal sense – but more or less a straightforward continuation of modern molecular biology, genetic research or genetic engineering, the substantial (concrete) application potential of synthetic biology raises the question whether the existing laws are sufficient or whether new laws are necessary.

In its most recent decision about the GenTG, the Federal Constitutional Court clarified that the legal authorities hold a special duty of care when evaluating the long term consequences of genetic engineering because the scientific state of knowledge has not yet been finalized. In this respect, the mandate of Art. 20a of the Basic Law (GG), which urges the legal authorities to accept their responsibility for future generations by protecting natural resources, must be observed. "This mandate demands for hazard control as well as for the prevention of risks. The environmental goods protected by Art. 20a GG include the preservation of biological variety as well as the protection of a species-appropriate life for endangered animal and plant species" (BVerfGE 2010). In this context, the Federal Constitutional Court made it clear that the aim of the regulations contained in the GenTG is particularly to guarantee protection from uncontrolled proliferation of genetically modified organisms. The legislature, however, has to take new knowledge and new scientific discoveries into account (European Commission 2000) and to check whether changes in risk assessment practices are necessary. If such is the case, the legislator must respond accordingly and adapt the legislation. Should the new risk level exceed the socially acceptable level of risk, the legislator needs to take action. The legal authorities have a duty to maintain a high level of safeguard, though not the highest possible level of protection of human health. If they do not fulfill this

---

2 It seems possible to achieve this object by appropriate interpretation of the law or by a decision of the ZKBS, in the same way as the ZKBS evaluates new GMOs or new forms of genetic engineering.

obligation, the jurisdiction may ultimately notice a violation of the precautionary principle (EuGH 2013).

2   Areas of Conflict

Irrespective of the question of how synthetic biology is defined, it should be noted that various ethical, social, and legal areas of conflict are the subject of the debate surrounding synthetic biology. As mentioned above, the debate focuses, on the one hand, on aspects of (bio-) safety/security.[3] The concept of biosafety is understood in terms of product and environmental safety, whereas biosecurity refers to protection against the abuse of synthetic biology, such as military or terrorist activities aided by the production of biological weapons. On the other hand, a very important focus of the debate concerns intellectual property rights (justice/fairness). These areas of conflict especially raise important legal questions.

2.1   Biosafety

As in the early development of recombinant DNA, many concerns have been raised in relation to synthetic biology (Balmer 2008, p.15). In 2004 the journal *Nature* called for a broad debate on the risks: "This is not only genes to be replaced. Now life is molded like clay. [...] The scope of such instruments is much larger than that of genetic modification and it is certainly much harder to foresee the actual risks" ("Futures of artificial life", 2004). Critics of synthetic biology maintain that the organisms can spread very quickly, because ecosystems are not prepared for their new properties (IRGC 2008). The European Group on Ethics (EGE) is concerned that: "Synthetic microorganisms released into the environment could initiate processes of horizontal gene transfer and affect biotic balances, or evolve beyond their functionality and elicit unprecedented side-effects on the environment and other organisms" (EGE 2009, pp.27f.). Further, "biological machines could evolve, proliferate and produce unexpected interactions that might alter the ecosystem" (EGE 2009, p.27). Therefore, adequate biological control must be developed, for example by "engineering bacteria to be dependent on nutrients with

---

3   The German language has only the single term 'Biosicherheit' to describe the area of conflict, whereas English distinguishes between biosecurity and biosafety. The English distinction will be used also in the following discussion.

limited availability, and integration of self-destruct mechanisms that are triggered should the population density become too great" (EGE 2009, p.27). Christoph Then from Testbiotech, a critical NGO, describes several risks of "the new methods for manipulation of the DNA which are here referred to Synthetic Genome Technologies" (Then 2013, own translation).

The risks posed by modified or completely new forms of life are complex. They do not only entail interaction with the environment; gene regulation in living things leaves room for effects that go far beyond additive effects of the individual components. Generally, the problem is that in a risk assessment of synthetic organisms only limited or no recourse to experience with existing life forms is possible (ZKBS 2012, p.8). Despite awareness of the potential risks, almost all publications on synthetic biology currently come to the conclusion that there is no need for legal investigation and action in the field of biosafety. The European Group on Ethics in Science and New Technologies (EGE), for example, states:

> At the moment virtually all approaches to synthetic biology involve the use of genetic modification techniques. Therefore, within the EU they are regulated through the Directives and Regulations for genetic modification introduced initially in 1990 and substantially modified during the ensuing years (EGE 2009).

The Joint Policy Paper by DFG, acatech and Leopoldina (2009) concludes that there is currently no, or at least not an essential, need to take action. The areas of conflict of synthetic biology are covered by existing law and therefore sufficiently regulated. The German government comes to the same conclusion (German Bundestag 2009; Donner et al. 2009). Questions relating to biosafety are covered by the Genetic Engineering Act (GenTG), the Medicinal Products Act (AMG), the Protection against Infection Act (IFSG), and the Chemicals Act (ChemG). According to DFG et al. (2009), these regulations are currently sufficient to the greatest possible extent, leaving no acute need to take immediate action. This is also the prevailing point of view shared by German authorities (Luttermann 2011, p.195; Friedrich 2010; Ethikbeirat 2009; Kupferschmidt 2011a).

The German Council of Ethics also sees no need for action, because synthetic biology in Germany falls completely within the scope of the GenTG, and aspects of biosafety are therefore mostly irrelevant. Currently, the most important task may be to establish a consistent definition of synthetic biology, to clearly distinguish it from other technologies, and to formulate an answer to the question of what the essential novelty of this technology actually is. The Council also acknowledges the fact that the development of synthetic biology may create new problems and security risks which demand a response, or rather, a debate on how to respond. For that reason, the importance of some form of monitoring process and its constant improvement are emphasized. This monitoring process must be improved constantly (Catenhusen 2011, pp.85-86). Monitoring is obligated by law. The ZKBS (Central Commission for Biological Safety) has already complied with the requirement for monitoring synthetic biology and has submitted an initial report on the topic (ZKBS 2012). In this report the ZKBS, in accordance with a duty of evaluation and observation assigned to it, investigates several new techniques that count as synthetic biology and finally comes to the conclusion that they either fall within the scope of the GenTG or – if this is not the case – do not create risks that require regulation. The same also applies to cells created de novo or to orthogonal biological systems.

NGOs such as Testbiotech with their leading protagonist Christopher Then, GeN,[4] and the BUND[5] advocate a much more critical position. They discern the need for legal action by the legislative authorities particularly with regard to the risks of synthetic biology for the environment in the case of setting free genetically modified organisms (a lack of options for control and recovery, lacking comparability of new organisms). The BUND proposes a moratorium for synthetic biology (Stagemann 2011, p.14). Testbiotech recommends a moratorium on research funding in the field of synthetic biology and also specific changes of the GenTG (Then and Hamberger 2010). According to Testbiotech, Section 1 GenTG should be supplemented as follows (Then 2010): "The aim of this law is [...] the environment [...] and the protection from an uncontrolled proliferation of genetically modified or synthetically produced organisms [...]."

Supplement of the legislative purpose as defined in Section 1 GenTG is merely declaratory and thus not necessary. In its most recent decision on the GenTG, the Federal Constitutional Court clarified that the aim of the regulations contained in the GenTG is, in particular, to guarantee protection from uncontrolled proliferation of genetically modified organisms (BVerfGE 2010).

Furthermore, Testbiotech demands that Section 16 GenTG be supplemented as follows: "(2) A release of genetically modified or synthetically produced organisms must be prohibited if their proliferation cannot be controlled or their retrieval is not ascertained."

In general, a regulation of this kind cannot be rejected from a constitutional point of view (Krämer 2013). Evaluation of the risk of endangerment falls within the prerogative of the legislative authority and does not require scientific-empirical proof of the real potential of endangerment due to genetically modified organisms and their progeny. In a situation that cannot be clarified scientifically, the legal authorities are entitled to evaluate dangers and risks, even more so since the pro-

---

4 Gen-ethisches Netzwerk e.V. (Gen-ethical Network).
5 Bund für Umwelt- und Naturschutz Deutschland e. V. (Friends of the Earth Germany).

tected legal goods are fixed in the constitution and have a high value, and since the existing risk of unwanted or harmful, and perhaps even irreversible, effects ought to be controlled with a view to greatest possible precaution. The Federal Constitutional Court (BVerfGE 2010) further refers to explanatory statements No. 4 and No. 5 to Directive 2001/18/EC (EC 2001).

Ultimately it will be almost impossible to provide conclusive evidence that unwanted proliferation of genetically modified or synthetically produced organisms can be controlled and that their retrieval is guaranteed in all cases. The supplementation of Section 16 GenTG postulated by Testbiotech would not only affect synthetic biology. It would in fact establish a ban on setting free genetically modified organisms into the environment for all areas of genetic engineering, i.e., for all genetically modified organisms (Then and Hamberger 2010). A restriction that has been postulated since the very first debates on genetic engineering would be put in place. Therefore, such a demand will hardly be politically acceptable. The call for such a restrictive regulation regarding the release of GMOs gives reason to speculate that the critics of genetic engineering are using the supposed novelty of synthetic biology in order to discuss and finally enforce their old demands for the limitation of genetic engineering (Eberbach 2012, pp.24ff.). This would probably spell the end of synthetic biology and genetic engineering in Germany. Even the long term investigation of environmental compatibility demanded by the EGE would hardly be possible because, ultimately, such a study would require the release of organisms. Only a controlled release of GMOs can provide 'real' and comprehensive findings on environmental compatibility in natural surroundings.

2.2 Biosecurity

In the U.S. it was possible to order different genetic DNA from more than 30 gene synthesis companies. The production and mailing of DNA parts, which could subsequently be combined to form, for example, the Marburg virus or smallpox, was completely legal. Such circumstance indicates one of the biggest risks posed by synthetic biology, namely the problem of biosecurity. The abuse of synthetic biology or misuse of gene synthesis to create new biological weapons by terrorists – collectively referred to as "biohacking" – is the main concern in the field of synthetic biology (Schultz 2009). Ethics experts of the European Commission explicitly warn against the risk of synthetic biology being abused to produce chemical weapons or for terrorist attacks. Similar fears have been voiced by the German Research Association (DFG et al. 2009).

There is concern that individuals, terrorist organizations or states have the option to reconstruct pathogenic organisms or toxins and use them for hostile or warlike acts. A similar approach could be of concern and prosecuted by persons who such as computer hackers and computer viruses designers, or as interested lay people receive access to individual synthetic elements or the necessary substances and produce synthetic systems in an uncontrolled environment, including micro-organisms (Schultz 2009).

The possible misuse of gene synthesis to create new biological weapons is openly discussed even among those engaged in synthetic biology (Garfinkel et al. 2007, pp.38ff.).

In Germany, the issues raised in the above mentioned papers are also being debated by the Ethics Council, the German Research Association, acatech et al. (in 2009) and the German government. The authors conclude that issues relating to biosecurity are covered by the "Biological and Toxin Weapons Convention" (BTWC: http://www.opbw.org/) as well as by other international (Wassenaar Arrangement: http://www.wassenaar.org/), European (EC 2009), and national regulations and provisions at international level, including the "Kriegswaffenkontrollgesetz" (War Weapons Control Act: KrWaffKontrG 2013).

In the international context, various international regimes are concerned with the issue of export control: the topic of chemical and biological weapons is dealt with by the 'Australia Group' (www.australiagroup.net). The 'Australia Group' has addressed the issue of synthetic biology on several occasions, in particular through its "New and Evolving Technologies Experts Group" (cf. BAFA 2012, p.28). At this level, there is some form of monitoring of synthetic biology, and the lists of dual-use goods and technologies, especially the Control List of Dual-use Biological Equipment Related Technology and Software, are frequently updated (Australia Group 2015). And the German government notes: "These legal regulations are supported by voluntary self-commitments of the industry to ensure that only trusted recipients get access to potentially abusive nucleic acids" (German Bundestag 2009, p.5).

Two developments that lately have been hotly debated in the context of synthetic biology can be highlighted: biohacking or do-it-yourself Biology and the restriction of freedom of research and scientific publication.

2.3 Biohacking

With regard to so-called do-it-yourself biology and the biohacker community that is currently also developing in Germany (Charisius et al. 2012), the German industry is calling for action and regulation (Schmidt 2011, p.120; Engelhard 2010, p.22). This demand concurs with that of the EGE to control and manage biohacking.

In connection with the subject of biohacking, institutional measures are proposed to prevent gene segments being synthesized from dangerous microorganisms and subsequently sold: labs should be protected (for example, through certification), gene synthesis devices registered, and the order of gene sequences controlled. Nouri and Chyba propose to equip the machines with built-in synthesis blocking mechanisms to make the synthesis of certain genes impossible (Nouri and Chyba 2009, pp.234-236). These proposals came from representatives of the two largest gene synthesis companies (DNA 2.0 and Geneart).

The above requirements appear reasonable. As the technical equipment and technical possibilities of the biohacking scene are certainly still limited, there is little real danger at present. Moreover, biohacking would fall within the scope of the Genetic Engineering Act if it should meet the factual requirements of the law. Nevertheless, precaution is advised. The technical possibilities can develop faster than currently envisaged. At the same time, the technical equipment will no doubt be available more readily and at a lower cost in future. At the beginning of the 1980s no one expected the rapid development of the internet as witnessed in recent years, or even foresaw the extent of abuse by computer hackers we find today. If provisions are not made now, the biohacking scene in future will likely be just as difficult to control as the computer hacker scene today.

2.4 Freedom of research and publication

At the end of 2011 an explosive issue arose when the blueprint for a deadly and highly contagious virus was going to be published. For several months, scientists and security experts debated the issue, and finally even the World Health Organization (WHO) intervened in the discussion. Some researchers claimed a 'Marshall Plan for biosecurity policy'. Pat Mooney, for example, argued that responsible scientists working together with civil society could trigger a movement for slow growth. Such a movement would need to agree in particular on an "International Convention for the Evaluation of New Technologies" (ICENT). Therefore, a code of ethics and standards should emerge for biological engineering as it has done for other engineering disciplines (Mooney 2010, p.118).

The WHO was the first official body to intervene, calling for full publication of all details of two studies by Yoshihiro Kawaoka and Ron Fouchier, respectively, on the H5N1 virus – though it only did so at a later date (WHO 2011). For the first time in the history of biological science, the U.S. National Science Advisory Board for Biosecurity (NSABB 2012) had recommended secrecy to protect mankind – also against bioterrorists. A blueprint for the super pathogens would be a monstrous weapon, the NSABB stated. Only a censored version of the research should be published.⁶ The NSABB was also considering a moratorium on similar cases.

In February 2014 a larger group of experts agreed in discussions with the WHO that the moratorium on experiments with the highly dangerous artificial bird flu virus, such as those conducted in high-security laboratories in Rotterdam, should be extended until further notice. The final document stated that it should be achieved in more public debates that security should be strengthened even more, just as the understanding of security of the population. And in 2012 two lawyers from Georgetown University in Washington had already clarified that the requirement of the U.S. health authorities to suppress technical information does not represent an abuse of power or illegal censorship and is certainly consistent with the U.S. Constitution (Kraemer and Gostin 2012).

Some researchers – especially in Germany – resist any approach that they perceive to limit freedom of research. "Such censorship is a clear breach of the rules of good scientific practice and leads to unnecessary dramatization," the German Society for Virology has stated. The influenza expert Hans-Dieter Klenk from Marburg University has also criticized the decision: "I would argue that research should be published in detail", he said. "A sensational conclusion cannot be easily set in the world. Research results will be published so that other researchers can repeat it and check. This is science" (Kupferschmidt 2011b, own translation).

In this context the question of the relationship between freedom of research and the precautionary principle arises (see, e.g., Würtenberger and Tanneberger 2014, pp.1-10). Is freedom of research unlimited in all cases? If it is not, to what extent should, or can, it be restricted by the authorities?

Freedom of research is a recognized element of the constitutional tradition of the Member States of the European Union. In Germany it is anchored in the constitution, in Article 5, paragraph 3, sentence 1 of the Basic Law (Grundgesetz): *"Art and science, research and teaching are free."*⁷ In the European Union it is anchored in Art. 13 of the EU Charta of Fundamental Rights (Jarass 2013). Academic freedom is guaranteed as an unconditional basic right, although the constitutional guarantee does not cover the state and society as such, but ultimately the welfare of the individual and the community serving science (BVerfGE 1978, 47,327,369). Like all other basic rights which are guaranteed without reservation, scientific freedom may be limited by reason of conflicting constitutional law (Cf. BVerfGE 1978, 47,327,369; 1981, 57,70,99). The protection of human life and health, freedom

---

6   Both studies have since been published (Imai et al. 2012; Fouchier 2012)
7   Own translation. Original: "Kunst und Wissenschaft, Forschung und Lehre sind frei." (GG 2014, art.5.1)

of profession and property of persons potentially concerned (Art. 2 para. 2 sentence 1; Art. 12 para. 1; Art. 14 para. 1 of the Basic Law), and the protection of natural resources (Art. 20a of the Basic Law) are important goods of constitutional rank which do justify a limitation of scientific freedom in the context of proportionality. A conflict between basic rights protected by the constitution shall be solved by recourse to further corresponding regulations and principles as well as to the principle of practical concordance by interpretation of the constitution (Cf. BVerfGE 1978, 47,327,369; 2008, 122,89,107). In the competition between conflicting, and also constitutionally protected values, freedom of research has no absolute priority. A tradeoff in the single case is necessary (VerfGH NRW 2000). Thus, for example, research objectives such as the development of nuclear and biological weapons can be excluded, or certain methods, such as experiments on humans, can be regulated (Security and Defense Research - Working Group 2010, pp.3f.). A complete ban on research on a specific topic or area is much more difficult (maybe even impossible) than to justify the rules of practicing research (Würtenberger and Tanneberger 2014, pp.6,8f.). Generally, there has to be a legal basis for imposing a ban of that kind (Cf. BVerfGE 1990, 83,130,142; 2003, 107,104,120; 2008, 122,89,107). In some cases, freedom of research can be limited by internal regulations. For example, in March 2010, the German Max Planck Society issued guidelines and rules concerning the responsible utilization of freedom of research and research risks.[8] The regulations were developed by the Security and Defense Research - Working Group (2010) with the support of the Ethics Council. The Max Planck Society also defined principles for responsible ethical research, which are binding for all employees of the Society (Security and Defense Research - Working Group 2010, pp.6f.). Similar principles or guidelines of other organizations could, in principle, be applied to many organizations. They could also be made internationally binding.

The subject of limitation of freedom of publication and freedom of science, which has been formulated at European level, has also found its way into the German discussion. Such postulations are hardly compatible with the German understanding of scientific freedom, which is protected by the German constitution. Freedom of research includes the evaluation of research results and their dissemination. Public discussion of research results is a central element of research. Therefore, individual scientists are, in principle, free to decide when and where their findings are published themselves (VG Köln 2012). However, the positive freedom of publication is not unlimited.

---

8   The Max Planck Society is very aware of the problem of 'dual-use' and also places emphasis on it.

A legal obligation of science to deal with possible consequences of the application and use of research results is also constitutionally legitimized. Such obligation also includes outward, observable, and verifiable knowledge-to-place and valuation processes (VerfGH NRW 2000).

Researchers may also be obliged to wait for the publication of their findings for a certain period of time (2 months) to allow their university to examine whether a safeguard rights management is concerned. To protect the universities rights of exploitation, even the framework can be staked out within the scientists may publish (BGH 2007). If a university wishes to exploit certain research findings a temporal restriction and / or restriction of the content of the publication is possible. In light of the above, the question is, whether a (restrictive) determination of criteria for the publication of data of highly pathogenic viruses or toxic substances should not be seen as proportionate and appropriate?

Before scientific freedom is limited by such a measure, it needs to be discussed whether more moderate, but equally effective means can be found to regulate synthetic biology, and biohacking in particular, with regard to the use of and access to pathogenic DNA sequences and technical instruments (sequencing instruments etc.) (German Bundestag 2009, p.5). This seems to be doubtful. Restrictions of freedom of publication have a lesser impact than a restriction of research itself. Publication restrictions do not impede research activity. Publication of the research results is even possible. The specific constitutional design of the restriction of freedom of publication is certainly not an easy task, but limitations of the freedom of publication can be justified constitutionally (Würtenberger and Tanneberger 2014, p.9; different opinion: Kluth 2012).

3 Synthetic Biology and Intellectual Property Rights/Patenting

The issue of intellectual property rights (IPR) is regularly associated with patents although the latter represent only one type of IPR, albeit the most important one (Colussi 2012, p.38). A first question regarding possible challenges to the patenting of synthetic biology inventions is whether the patent process for synthetic biology is essentially different to the current patenting system, and secondly, whether the traditional patenting system is effective enough to handle the new developments. With regard to current patent law in general, we can note that the patentability of microorganisms and higher life forms, including genetically modified organisms, has been confirmed under the European Patent Convention and its case law (Rutz

2009, p.14). Thus, this is not a special problem for synthetic biology ('essence of life' issue) (Colussi 2012, p.37).

Following Rutz, we can also state that any difference lies in the scale of modification of naturally occurring organisms with tens to hundreds of genes by synthetic biology, and in the underlying concepts based on an engineering approach or the creation of artificial life, rather than in the technologies used. Many of the buildings blocks or 'parts' are identical to those used in other areas of biotechnology (Rutz 2009, p.15). Current patent law will therefore be applicable in general. Nevertheless, the complex character of synthetic biology with its interdisciplinarity, complexity, interconnectedness, interoperability, and standardization could lead to many problems in practice (Rutz 2009, p.15). As a result it could become more difficult to apply for patents for new inventions made by synthetic biology. Ultimately, a situation could arise where a synthetic biology patent is composed of many other patents and other components, involving hundreds of different parts – creating a 'patent thicket'.

It is doubtful whether cross licenses among patent holders, patent pools or clearing houses can provide an effective solution. It seems all the more doubtful considering the fact that patent thickets are targeted by patent trolls or patent sharks for the purpose of suing other companies. The situation could become totally confusing for patent applicants. It will not suffice to simply maintain or raise the bar of patentability, to require clear boundaries, or to ensure that the patenting process is transparent and that patents are easily detectable and retrievable by all those working in the field. While such measures would no doubt be helpful, they cannot provide an effective solution.

Therefore, the most important recommendation is to reduce complexity in conjunction with the adoption of other useful strategies such as those proposed by Henkel and Reitzig (2008) (Miguel Beriain 2013). These strategies include: not building huge patent portfolios for the purpose of cross licensing among competitors; simplifying technical standards and creating more modular designs for companies; better interdepartmental and inter-company cooperation; and early cooperation with competitors in the R&D process. Finally, companies should stop flooding patent offices with insignificant inventions, and public authorities should support and actively intervene in the patent process to foster cooperation. Patenting could then become efficient and support the commercialization of synthetic biology.

4 Summary

In summary, it can be stated that synthetic biology is largely equated with biotechnology in the jurisprudential discussion, especially with regard to genetic engineering, and thus is nothing fundamentally new. An intensive jurisprudential debate on synthetic biology does not take place. Thus, only a handful of original jurisprudential publications on synthetic biology exists. According to prevailing opinion, there is currently no need to take action and set up regulations. The same applies to the issue of biosecurity, even if many experts warn about the dangers arising from synthetic biology especially in this area. The only aspect to have been acknowledged is the necessity of broad observation and discussion of synthetic biology, as postulated by many in Germany, Europe, and the United States of America. The debate on areas of conflict within the field of synthetic biology is still in an early phase of orientation. Nonetheless, it is imperative that the interdisciplinary approach, which is typical for synthetic biology and encompasses different technical and natural scientific disciplines, be enlarged to include the humanistic disciplines with the aim of facilitating a holistic debate.

References

Australia Group (2015). *Control List of Dual-use Biological Equipment and Related Technology and Software.* http://australiagroup.net/en/dual_biological.html. Accessed: 4 May 2015.
BAFA (2012). Bundesamt für Wirtschaft und Ausfuhrkontrolle. Annual report of the Federal Office of Economics and Export Control 2012/2013. http://www.bafa.de/bafa/de/das_bafa/publikationen/das_bafa_bericht_2012_2013.pdf. Accessed: 11 May 2015.
Balmer, A., & Martin, P. (2008). Synthetic Biology - Social and Ethical Challenges. Nottingham: Institut for Science and Society, University of Nottingham.
BGH (2007). Bundesgerichtshof. German Federal Court Judgment of 18 September 2007 – X ZR 167/05.
Boldt, J., Müller, O., & Maio, G. (2009). *Synthetische Biologie. Eine ethisch-philosophische Analyse* (Vol. 5). Bern: BBL.
BVerfGE (1978). Bundesverfassungsgerichtsentscheid. Adjudication of the German Constitutional Court of 1 March 1978. In: BVerfGE 47,327.
BVerfGE (1981). Bundesverfassungsgerichtsentscheid. Adjudication of the German Constitutional Court of 8 April 1981. In: BVerfGE 57,70.
BVerfGE (1990). Bundesverfassungsgerichtsentscheid. Adjudication of the German Constitutional Court of 27 November 1990. In: BVerfGE 83,130.
BVerfGE (2003). Bundesverfassungsgerichtsentscheid. Adjudication of the German Constitutional Court of 16 January 2003. In: BVerfGE 107,104.

Legal Aspects of Synthetic Biology

BVerfGE (2008). Bundesverfassungsgerichtsentscheid. Adjudication of the German Constitutional Court of 28 October 2008. In: BVerfGE 122,89.

BVerfGE (2010). Bundesverfassungsgerichtsentscheid. Adjudication of the German Constitutional Court of 24 November 2010. In: BVerfGE 128,1.

Catenhusen, W.-M. (2011). Schlusswort. In: *Werkstatt Leben. Bedeutung der Synthetischen Biologie für Wissenschaft und Gesellschaft* (pp.84ff.). Notes of the public conference of 23 November 2011. http://www.ethikrat.org/dateien/pdf/tagung-23-11-2011-simultanmitschrift.pdf. Accessed: 7 May 2015.

Charisius, H., Friebe, R., & Karberg, S. (2012). Biotechnologie: Unser kleines Gen-Labor. *Spektrum der Wissenschaft*. http://www.spektrum.de/news/unser-kleines-gen-labor/1153300. Accessed: 7 May 2015.

Colussi, I.A. (2012). *Synthetic biology, concerns and risks: looking for a (constitutionally oriented) regulatory framework and a system of governance for a new emerging technology*. PhD Thesis. Trent: University of Trent. http://eprints-phd.biblio.unitn.it/966/2/Colussi_I.A.-Ph.D.Thesis.pdf. Accessed: 7 May 2015.

Dederer, H.-G. (2010). Neuartige Technologien als Herausforderung an das Recht – dargestellt am Beispiel der Nanotechnologie. In: T.M. Spranger (ed.), *Aktuelle Herausforderungen der Life Sciences* (pp. 71f.). Münster: LIT Verlag.

DFG, acatech, & Leopoldina (2009). Deutsche Forschungsgemeinschaft (German Research Foundation), acatech – Deutsche Akademie der Technikwissenschaften (German academy of technological sciences), Leopoldina – Deutsche Akademie der Naturforscher (German academy of natural scientists). *Synthetic Biology: Positions*. Weinheim: Wiley VCH.

Donner, S., & Winter, A. (2009). Synthetische Biologie. Statement of the Research Section (Wissenschaftliche Dienste, WD) of the German Bundestag, No. 60/09. https://www.bundestag.de/blob/190720/bc605aca84bda4a5f6610ff4b68104cd/synthetische_biologie-data.pdf. Accessed: 7 May 2015.

Eberbach, W. (2012). Gentechnik und Recht. In: W. Eberbach, P. Lange, & M. Ronellenfitsch (eds.), *Recht der Gentechnik und Biomedizin*, 79. Ergänzungs-Lieferung, Band 1, Teil A. I. Heidelberg: C.F. Müller.

EC (2001). European Council. DIRECTIVE 2001/18/EC OF THE EUROPEAN PARLIAMENT AND OF THE COUNCIL of 12 March 2001 on the deliberate release into the environment of genetically modified organisms and repealing Council Directive 90/220/EEC, OJ 2001 L 106/1.

EC (2009).European Council. COUNCIL REGULATION (EC) No 428/2009 of 5 May 2009 setting up a Community regime for the control of exports, transfer, brokering and transit of dual-use items, OJ 2009 L 134/1.

EGE (2009). European Group on Ethics. The European Group on Ethics in Science and New Technologies to the European Commission. Ethics of synthetic biology, Opinion No. 25, Brussels, 17. November, p.27f.

Engelhardt, M. (2010). Biosicherheit in der Synthetischen Biologie – Die Unterschiede zur Gentechnik erfordern neue Sicherheitsstandards. *Die Politische Meinung*, 493, 17-22.

Ethikbeirat (2009). *Bericht über die Arbeit des Ethikbeirates*. BT-Drucksache 16/13780 of 1 July 2009.

EuGH (2013). Europäischer Gerichtshof. Judgement of the Court of Justice of the European Union of 11 July 2013, C-601/11P. ECLI:EU:C2013:465.

European Commission (2000). Communication from the Commission on the precautionary principle of 2 February 2000, COM(2000) 1. http://ec.europa.eu/dgs/health_consumer/library/pub/pub07_en.pdf. Accessed: 4 May 2015.

Fink, G. R., Atlas, R., Barkley, W. E., Collier, R. J., Cozzens, S. E., & Faden, R. (2004). *Biotechnology Research in an Age of Terrorism.* Washington, DC: The National Academies Press.

Fouchier, R.A. (2012). Airborne transmission of influenza A/H5N1 virus between ferrets. *Science, 336*(6088), 1534-41.

Friedrich, B. (2010). Talk at the Forum Bioethik *Synthetische Biologie – Leben aus dem Baukasten?* of 24. February 2010. Deutscher Ethikrat. Audio transcript: http://www.ethikrat.org/dateien/audio/fb_10-02-24_03_vortrag.mp3. Accessed: 7 May 2015.

Futures of artificial life (2004). *Nature, 431*(7009), 613.

Garfinkel, M.S., Endy, D., Epstein, G.L., & Friedman, R.M. (2007). *Synthetic Genomics. Options for Governance.* J. Craig Venter Institute, CSIS, MIT.

German Bundestag (2009). *Stand und Perspektiven der Synthetischen Biologie.* BT-Drucksache 17/5165 of 22 March 2011.

GG (2014). Grundgesetz für die Bundesrepublik Deutschland (German Basic Law) in der im Bundesgesetzblatt Teil III, Gliederungsnummer 100-1, veröffentlichten bereinigten Fassung, das zuletzt durch Artikel 1 des Gesetzes vom 23. Dezember 2014 (BGBl.I S. 2438) geändert worden ist.

Henkel, J., & Reitzig, M. (2008). Patent sharks. *Harvard Business Review, 86*(6), 129.

Herbig, J. (1978). *Die Gen-Ingenieure: Durch Revolutionierung der Natur zum neuen Menschen?* München, Wien: Hanser.

Herdegen, M. (2012). Einleitung GenTG. In: W. Eberbach, P. Lange, & M. Ronellenfitsch (eds.) (2012), *Recht der Gentechnik und Biomedizin,* Band 1, Teil I (p.11). Heidelberg: C.F. Müller.

Hohlfeld, R. (1984). Der Mensch als Objekt von Biotechnologie und biomedizinischer Forschung. *Gewerkschaftliche Monatshefte, 35,* 594-596.

Hohlfeld, R. (1986). Die zweite Schöpfung des Menschen. *Gewerkschaftliche Monatshefte 9,* 550-560.

Hohlfeld, R. (1988). Biologie als Ingenieurskunst. Zur Dialektik von Naturbeherrschung und synthetischer Biologie. *Ästhetik und Kommunikation, 69,* 61ff.

Hohlfeld, R. (1990). Synthetische Biologie – Biologie als Ingenieurskunst. In: K. Grosch (ed.), *Herstellung der Natur? Stellungnahmen zum Bericht der Enquete-Kommission „Chancen und Risiken der Gentechnologie".* Frankfurt/Main: Campus.

Imai, M., Watanabe, T., Hatta, M., Das, S.C., Ozawa, M., Shinya, K., ... Kawaoka, Y. (2012). Experimental adaptation of an influenza H5 HA confers respiratory droplet transmission to a reassortant H5 HA/H1N1 virus in ferrets. *Nature, 486*(7403), 420-428. doi: 10.1038/nature10831.

Jarass, H.D. (2013). *Charta der Grundrechte der Europäischen Union.* München: Beck.

Kluth, W. (2012). *Wissenschaftsfreiheit vs. Sicherheitsinteressen.* http://www.academics.de/wissenschaft/wissenschaftsfreiheit_vs_sicherheitsinteressen_52504.html. Accessed: 7 May 2015.

Kraemer, J.D., & Gostin, L.O. (2012). The Limits of Government Regulation of Science. *Georgetown Law Faculty Publications and Other Works.* Paper 776. http://scholarship.law.georgetown.edu/facpub/776. Accessed: 7 May 2015.

Krämer, L. (2013). Genetically Modified Living Organisms and the Precautionary Principle. Testbiotech e.V. dossier. http://www.testbiotech.org/sites/default/files/GMO%20and%20precaution.pdf. Accessed: 7 May 2015.

Legal Aspects of Synthetic Biology

KrWaffKontrG (2013). Kriegswaffenkontrollgesetz (War Weapons Control Act). Gesetz über die Kontrolle von Kriegswaffen in der Fassung der Bekanntmachung vom 22. November 1990 (BGBl. I S. 2506), das zuletzt durch Artikel 2 Absatz 2 des Gesetzes vom 6. Juni 2013 (BGBl. I S. 1482) geändert worden ist.

Kupferschmidt, K. (2011a). Ingenieure des Lebens (5): Gefahr aus dem Labor. *Der Tagespiegel*, 7 March 2011. http://www.tagesspiegel.de/wissen/ingenieure-des-lebens-5-gefahr-aus-dem-labor/3920994.html. Accessed: 7 May 2015.

Kupferschmidt, K. (2011b). Vogelgrippe: Mit den Waffen der Wissenschaft. *Der Tagespiegel*, 31 December 2011. http://www.tagesspiegel.de/wissen/vogelgrippe-mit-den-waffen-der-wissenschaft/6007782.html. Accessed: 7 May 2015.

Lengeler, J. (1988). Die methodischen Grundlagen der modernen Gentechniken. In: J. Baltzer (ed.), *Gentechniken und Individuum* (p.17). Köln: Heymann.

Luttermann, C. (2011). Synthetische Biologie: Bausteine für Leben und Jurisprudenz. In: *Juristenzeitung*, 66, 195f.

Miguel Beriain, I. d. (2013), Synthetic Biology and IP Rights. In: Defense Of The Patent System, Presentation Workshop *Engineering Life – Zur ethisch-gesellschaftlichen Relevanz der Synthetischen Biologie*, September 26-28, University of Freiburg. See also "Synbio and IP rights: looking for an adequate balance between private ownership and public interest" in this volume.

Mooney, P. (2010). *Next Bang! Wie das riskante Spiel mit Megatechnologien unsere Existenz bedroht*. München: Oekom.

Müller-Röber, B. (2011). Panel discussion. In: *Werkstatt Leben. Bedeutung der Synthetischen Biologie für Wissenschaft und Gesellschaft* (pp.70f.). Notes of the public conference of 23 November 2011. http://www.ethikrat.org/dateien/pdf/tagung-23-11-2011-simultanmitschrift.pdf. Accessed: 7 May 2015.

Nouri, A., & Chyba, C.F. (2009). Proliferation-resistant biotechnology: an approach to improve biological security. *Nature Biotechnology, 27*, 234-236.

NSABB (2012). National Science Advisory Board for Biosecurity. Meeting of the National Science Advisory Board for Biosecurity to Review Revised Manuscripts on Transmissibility of A/H5N1 Influenza Virus, Statement of the NSABB. http://osp.od.nih.gov/sites/default/files/resources/NSABB_Statement_March_2012_Meeting.pdf Accessed: 27 December 2012.

Ronellenfitsch, M. (2004). §3 GenTG. In: W. Eberbach, P. Lange, & M. Ronellenfitsch (eds.) (2012), *Recht der Gentechnik und Biomedizin*, Band 1, Teil I, B, I (pp.22f.). Heidelberg: C.F. Müller.

Ronellenfitsch, M. (2008). In: W. Eberbach, P. Lange, & M. Ronellenfitsch (eds.) (2012), *Recht der Gentechnik und Biomedizin*, Band 1, Teil I, B, I (p.17). Heidelberg: C.F. Müller.

Rutz, B. (2009). Synthetic Biology and patents. *EMBO Reports, 10*(S1), 14-17.

Sauter, A. (2011). Synthetische Biologie: Finale Technisierung des Lebens – oder Etikettenschwindel? *TAB-Brief Nr. 39*, 16-23.

Schmidt, M. (2011). Biosicherheit und Synthetische Biologie. In: A. Pühler, B. Müller-Röber, M.-D. Weitze (eds.), *Synthetische Biologie – Die Geburt einer neuen Technikwissenschaft* (pp.111-127). Berlin, Heidelberg: Springer.

Schmidt, M., & Giersch, G. (2011). DNA synthesis and security. In: M. J. Campbell (ed.), *DNA Microarrays, Synthesis and Synthetic DNA* (pp.285-300). Hauppauge NY: Nova Science Publishers, Inc.

Schultz, N. (2009). *Perspektivenpapier Synthetische Biologie*. http://www.ethikrat.org/dateien/pdf/Perspektivenpapier_Synthetische_Biologie_2009-04-23.pdf. Accessed: 11 May 2015.

Schummer, J. (2011). *Das Gotteshandwerk: Die künstliche Herstellung von Leben im Labor*. Berlin: Suhrkamp.
Schwille, P. (2011). Synthetische Biologie – Konstruktionsansätze für Lebensprozesse? In: *Werkstatt Leben. Bedeutung der Synthetischen Biologie für Wissenschaft und Gesellschaft* (pp.5ff.). Notes of the public conference of 23 November 2011. http://www.ethikrat.org/dateien/pdf/tagung-23-11-2011-simultanmitschrift.pdf. Accessed: 7 May 2015.
Security and Defense Research - Working Group (2010). *Guidelines and Rules of the Max Planck Society On A Responsible Approach To Freedom Of Research And Research Risks*. https://www.mpg.de/232129/researchFreedomRisks.pdf. Accessed: 17 November 2012.
Stegemann, R. (2011). Contribution. In: *Werkstatt Leben. Bedeutung der Synthetischen Biologie für Wissenschaft und Gesellschaft* (p.14). Notes of the public conference of 23 November 2011. http://www.ethikrat.org/dateien/pdf/tagung-23-11-2011-simultanmitschrift.pdf. Accessed: 7 May 2015.
Then, C. (2010). Testbiotech fordert Schutz vor Risiken der synthetischen Biologie. Press release by TestBiotech e.V. of 15 June 2010. http://www.testbiotech.org/node/390. Accessed: 11 May 2015.
Then, C. (2013). Synthetic Genome Technologies. Testbiotech Background of 18 November 2013. http://www.keine-gentechnik.de/fileadmin/pics/Informationsdienst/Pflanzen/2013_11_18_Testbiotech_Synthetic_Genome_Technologies.pdf. Accessed: 11 May 2015.
Then, C., & Hamberger, S. (2010). *Synthetische Biologie. Teil 1: Synthetische Biologie und künstliches Leben – Eine kritische Analyse*. Testbiotech. https://www.testbiotech.org/sites/default/files/Synthetische%20Biologie%20Teil%201_7.Juni%202010.pdf. Accessed: 11 May 2015.
VerfGH NRW (2000). Verfassungsgerichtshof Nordrhein-Westfalen. Judgement of the Constitutional Court of North Rhine-Westphalia, Germany of 25 January 2000. VerfGH 2/98.
VG Köln (2012). Verwaltungsgericht Köln. Judgement of the Cologne Administrative Court of 6 December 2012. 13 K 2679/11.
WHO (2011) World Health Organisation. *WHO concerned that new H5N1 influenza research could undermine the 2011 Pandemic Influenza Preparedness Framework*. Statement. http://www.who.int/mediacentre/news/statements/2011/pip_framework_20111229/en/. Accessed: 11 May 2015.
Würtenberger, T., Tanneberger, S. (2014). Biosicherheit und Forschungsfreiheit. Zu den Schranken des Art. 5 Abs. 3 S. 1. GG. *Ordnung der Wissenschaft, 1/2014*, 1-10. http://www.ordnungderwissenschaft.de/pdf/2014-1/01_01_wuertenberger_tanneberger_biosicherheit.pdf. Accessed: 11 May 2015.
Zelder, O. (2011). Contribution In: *Werkstatt Leben. Bedeutung der Synthetischen Biologie für Wissenschaft und Gesellschaft* (p.15). Notes of the public conference of 23 November 2011. http://www.ethikrat.org/dateien/pdf/tagung-23-11-2011-simultanmitschrift.pdf. Accessed: 7 May 2015.
ZKBS (2012). Zentrale Kommision für die Biologische Sicherheit (Central Commission for Biological Safety and Security). *Monitoring der Synthetischen Biologie in Deutschland. 1. Zwischenbericht der Zentralen Kommission für die Biologische Sicherheit vom 6. November 2012*. http://www.bvl.bund.de/SharedDocs/Downloads/06_Gentechnik/ZKBS/01_Allgemeine_Stellungnahmen_deutsch/01_allgemeine_Themen/Synthetische_Biologie.pdf?__blob=publicationFile&v=3. Accessed: 26 March 2015.

Synbio and IP rights: looking for an adequate balance between private ownership and public interest

Iñigo de Miguel Beriain

1 Introduction[1]

In November 2009, a group of leading scientists, including two Nobel Prize winners, John Sulston and Joseph Stiglitz, published The Manchester Manifesto (2009), which constituted the conclusion of a long term effort to reflect on the ownership of science and the best way to manage it. The Manifesto was based on the idea that the Intellectual Property Rights system "has significant drawbacks in terms of its effects on science and economic efficiency, and raises ethical issues because of its (often adverse) effects on people and populations." Moreover, it stated that

> [i]t is clear that the dominant existing model of innovation, while serving some necessary purposes for the current operation of innovation, also impedes achievement of core scientific goals in a number of ways. In many cases it restricts access to scientific knowledge and products, thereby limiting the public benefits of science; it can restrict the flow of information, thereby inhibiting the progress of science; and it may hinder innovation through the costly and complicated nature of the system. Limited improvements may be achieved through modification of the current IP system, but consideration of alternative models is urgently required (The Manchester Manifesto 2009).

These brief paragraphs perfectly highlight the key points that are continuously at stake in all discussions about science and intellectual property rights: morals, efficiency, legitimacy, etc. — synthetic biology poses no exception. Furthermore, this

---

1   This paper is directly related to the research project "Synbio, Análisis de las implicaciones de la biología sintética en el ámbito de la propiedad intelectual (Código S-PE12UN013), SAIOTEK 2012 Departamento de Industria, Innovación, Comercio y Turismo del Gobierno Vasco" and the COST ACTION IS1001 "Bio-objects and their boundaries: governing matters at the intersection of society, politics, and science."

groundbreaking new discipline, which has potential for impressive applications in fields directly related to human health, such as biomedicine, seems to be stirring a debate between those willing to rule it through the patent based system and those who would much prefer at least to try to consider different options.

Over the next pages I will endeavor to uncover the rationalities that can be applied to synthetic biology in relation to human health. In doing so, I will seek to defend one basic idea: there is a need to combine all the tools provided by our current IP legal framework so as to achieve optimum development of this technology. However, before turning to that point, it is necessary to point out the different challenges that synbio poses in terms of IP issues.

2 IP and Synbio. Exploring the issues

The current IP rights protection system is based on the concept of the patent. As is commonly known, a patent

> protects new inventions and covers how things work, what they do, how they do it, what they are made of and how they are made. If a patent application is granted, it gives the owner the ability to take a legal action under civil law to try to stop others from making, using, importing or selling the invention without permission (Sapience, n.d.).

Patents therefore provide their owners the right to bar the use of an invention for a long term (usually twenty years), a bar that can only be avoided by paying a fee to obtain a license. Such measure may, traditionally, have been extremely useful in stimulating the development of science. Nonetheless, new technologies have raised a number of issues which seem to defy general agreement on the utility of patents. In the following section I will look at some of these issues and discuss them.

2.1 The moral issues

There are at least two relevant moral issues related to patents. Firstly, some people believe that the idea of private ownership of science, as such, is inacceptable. This radical point of view is extremely difficult to defend in as far as it seems to be contrary to Article 27 of the Universal Declaration of Human Rights (1948), which states that "[e]veryone has the right to the protection of the moral and material interests resulting from any scientific, literary or artistic production of which he is the author." Keeping this in mind, a less significant number of people consider that private interest must be put aside when public interest is at stake. This abstract idea usually takes concrete form in recommendations to reduce the length of patents or in the utilization of public licenses so as to afford the public access to knowledge when necessary. However, even this softer version of this claim seems difficult to achieve in a market controlled by the pharmaceutical industry, which is constantly under pressure due to decreasing financial margins. Initiatives of this kind might lead to a severe reduction in the amount of money dedicated to basic research, a matter that is at present extremely important for synbio.[2]

A different moral objection is held by those who focus on the subject matter of the patent instead of on the nature of private property rights. In their opinion, synbio ought to be banned whenever it involves human genetic material, human embryos, and so on. But this would surely also imply a rejection of patent claims, according to what is stated in Art. 6.2 of Directive 98/44/EC of the European Parliament and of the Council of 6 July 1998 on the legal protection of biotechnological inventions (EC 1998). This second objection makes more sense, considering that special protection of human genetic material has been recognized by the UN in its Universal Declaration on the Human Genome and Human Rights (1997, art.1), which states that "[t]he human genome underlies the fundamental unity of all members of the human family, as well as the recognition of their inherent dignity and diversity. In a symbolic sense, it is the heritage of humanity."

Whatever this means, we seem to have a solid argument for considering that certain kinds of material – such as human genes – should never be patented. At the same time, a crucial point is that the product, not the technique, should be definitive in this sense: for instance, a human embryo created by synbio should never be patented because it is a human embryo, not because it was created through that technique. Consequently, it seems the emergence of synbio *per se* should not make any difference in the intellectual property defense system, that is to say: if the moral clause has to be modified, it will not depend on synbio.

---

2   As Rai and Boyle (2007) stated: "While taking a drug all the way through clinical trials mandated by the US Food and Drug Administration may not cost as much as drug companies claim, it does cost hundreds of millions of dollars. Whether patent rights are the best incentive mechanism for purposes of eliciting pharmaceutical R&D is not a question we can address here. Suffice it to say that our current system of financing pharmaceutical innovation relies heavily on these rights".

2.2 The technical issues

Another prominent objection to the patenting system in relation to synbio does not concern moral issues, but the technical complexity of this new technology. As Rutz (2009) has stated,

> Synthetic biology often combines approaches from various disciplines, such as biotechnology, chemistry, nanotechnology, computing and engineering. This requires patent offices and patent professionals to be proficient in all of the technical domains involved. It puts a high burden on recruitment and training to ensure that the relevant prior art is found and judged correctly. Moreover, synthetic biology involves two areas, biotechnology and computing, both of which are at the centre of the public debate about patents. Some observers have therefore predicted a 'perfect storm' with regard to patents in synthetic biology.

The patent system was created at a time when it was very easy to solve the issues relating to non-obviousness, novelty, or practical application. Nowadays, biotechnology has raised the bar of complexity much higher. The problem does not necessarily concern the incapability of the officers working at the Patent Offices to adequately examine a patent's claim, but rather, the time and cost that this would involve. Nonetheless, determining different thresholds for what constitutes the key factors of patentability might give rise to severe dissension between different Offices. The higher the complexity, the greater is the scope for disagreement and dissention. Finally, it seems almost needless to mention that complexity increases uncertainty, in that it is rather difficult to anticipate whether one's patent application will be successful or not when a lot of different factors might play a role in the final decision. All these factors might be crucial in crushing middle-size biotech firms, thus promoting the creation of huge monopolies, which is something we should always worry about.

2.3 The utility issues

Last but not least, the most important criticism of the patent system in the case of synbio is that, due to the inherent complexity of this technology, the patent mechanism will not be able to stimulate its development and will provoke a general impediment of the whole market. This argument is directly related to what is usually called the 'tragedy of the anticommons', a type of coordination breakdown, where a single resource has numerous rightsholders who prevent others from using it, frustrating what would be a socially desirable outcome. Although unusual in the pharmaceuticals industry, such a scenario might be extremely common in the case of a technology built on the concept of aggregation. In this case, intellectual property rights will often be hard to identify because they remain fragmented across many owners, and even across many different countries, considerably augmenting the difficulties inherent to the complex nature of synbio. Were the LEGO metaphor still applicable, we would only need to imagine the difficulty of building a LEGO car when all the pieces belong to different children. In the field of electronics, where the fragmentation of property is extremely common, companies have traditionally found ways to avoid a general stalemate in the market by creating tools such as patent thickets and also in part by a mindset that always tries to avoid a general stalemate. It is difficult to predict whether pharmaceutical companies, which are much less used to cooperating with each other, will be able to reproduce this mindset in the case of synbio when it relates to human health.

The problem is especially relevant since complex biotechnologies have stimulated the creation of companies that do not produce anything, but earn profit through the complexity of the IP system. I am referring to 'patent sharks' or 'patent trolls', companies who have hidden intellectual rights and threaten to sue research and development (R&D) companies when their rights have been inadvertently infringed. These kind of parasite entities are eminently capable of considerably increasing the costs related to research and investment in the case of synbio.[3] In fact, the situation is so concerning that the president of the USA has recently announced his intention to fight them by introducing certain modifications in the current legal framework.[4]

---

3   Some data included in Wikipedia: "In 2011, United States business entities incurred $29 billion in direct costs because of patent trolls. Lawsuits brought by 'patent assertion companies' made up 61% of all patent cases in 2012, according to the Santa Clara University School of Law. From 2009 through mid-2013, Apple Inc. was the defendant in 171 lawsuits brought by non-practicing entities (NPEs), followed by Hewlett-Packard (137), Samsung (133), AT&T (127), and Dell (122). Patent troll-instigated litigation, once mostly confined to large companies in patent-dependent industries such as pharmaceuticals, came to involve companies of all sizes in a wide variety of industries. In 2005 patent trolls sued 800 small firms (those with less than $100 million annual revenue), the number growing to nearly 2,900 such firms in 2011; the median defendant's annual revenue was $10.3 million" (Patent troll 2015).

4   President Obama publicly addressed the issue of patent trolls on June 4, 2013 and directed the United States Patent and Trademark Office (USPTO) to take five new actions to help stem the surge in patent-infringement lawsuits tying up the court system. Saying "they don't actually produce anything themselves, they're just trying to essentially leverage and hijack somebody else's idea and see if they can extort some money out of them," Mr. Obama ordered the USPTO to require companies to be more specific about exactly what their patent covers and how it is being infringed. The Administration also told the patent office to tighten scrutiny of overly broad patent claims and said it would aim to

However, it is hard to tell whether it will be possible to respond effectively to the threat these entities pose.

Thus, I must conclude that patent systems undeniably have the potential to slow growth in the industry if firms abuse them. Besides, this will happen much more easily if complexity rises. But even then, however, it seems that we might possess certain legal and policy tools to avoid the worst thinkable scenario and promote, in contrast, the reasonable use of patents in combination with sui generis, or custom-made, intellectual property regimes that have been proposed as an alternative. In the last part of this paper I will endeavor to determine the best approach to adopt.

3 Stimulating cooperation, avoiding stalemate: time for politics and laws

The conclusion to be drawn from the arguments outlined above must be that the future of synbio will probably depend on its complexity and on the way stakeholders decide to cope with its complexity. Legal tools cannot play a role in the complexity issue, but they do play a decisive part in the cooperative issue. In fact, I am convinced that sound politics in synbio must seek to stimulate cooperation and avoid any risk of general stalemate. To that end, there are a number of initiatives to choose from, and various legal tools that perfectly fit that goal. Let me analyze some of these.

3.1 Public property, compulsory licensing

Promoting public property is surely one of the best ways to avoid a general obstruction of the system and to provide free general access to science and technology. The only trouble is that this could only happen in one of two ways. The first involves spending a huge amount of public resources on research programs. However, this approach generates at least two huge problems: 1) It is extremely expensive and provokes the substitution of private initiative with public-domain efforts, which is not always efficient. 2) Most of the currently used parts are already protected by IP rights. In theory, this inconvenience could be overridden through the use of

---

curb patent-infringement lawsuits against consumers and small-business owners who are simply using off-the-shelf technology. The President further asked Congress to pass laws that would have an even greater impact on curbing "abusive" lawsuits (Patent troll 2015).

the compulsory licensing mechanism.[5] However, that would definitively ruin any kind of private investment in this market. Consequently, in my opinion we need to carefully avoid taking radical measures of this sort in such a sensitive market as synbio seems to be.

The above statement certainly does not imply that we should not make use of these kinds of tools in some way. For instance, it seems to make sense to link public grants for private research to a promise of enabling the social use of IP rights derived from that research. It will also be a good idea to encourage the speedier development of this market by funding the kind of basic research that seems to be necessary but less promising in terms of returns on costs. With regard to compulsory licensing, it makes sense to support the use of this clause whenever circumstances require. The general obstruction of a major part of synbio research relating to human health as a result of a single patent might, at least in my opinion, perfectly justify this kind of intervention. However, it is very difficult to imagine that such a circumstance could emerge in practice.

3.2 Limiting the field of IP rights, avoiding trivial patents

A considerable number of authors have argued that the current IP framework could be significantly improved through policies that tighten the requirements for granting patents ('raising the bar policies'). This argument is based on the idea that it is particularly difficult to trace boundaries between what is obvious and what is not, or what is new and what is not, and sometimes Patent Offices seem to be too permissive. This is especially true in the case of the USA Patent Office when dealing with DNA patents, which has led to the situation where the number of such patents approved by the EPO is only 1/7th of the American total. The main difference here is the much wider interpretation of the idea of 'novelty' in the USA compared to Europe. Raising the bar would prevent companies from abusing the patent mechanism, the latter being one of the main reasons for the growing cost of technology. Moreover, it seems absolutely necessary to limit IP rights to the uses of the invention described in the patent claim, and to exclude any other use that might be described afterwards. As Rutz (2009) stated,

> [a]t least three responses on the part of patent offices and regulators can be envisaged to counteract these trends: first, to maintain or even raise the bar of patentability,

---

5   Let us bear in mind that the Doha Declaration allows countries to override patent laws to protect public health through compulsory licensing.

that is, to make sure that no patents are granted for obvious or even trivial subject matter; second, to make sure that patent rights have clear boundaries and are examined without much delay to provide legal certainty; and third, to ensure that patents and the patenting process are transparent, and easily detectable and retrievable by everybody working in the field.

However, this suggestion raises some important issues, too. The most obvious one is that the inclusion of concepts such as the 'trivial patent' would create an area of uncertainty, which is something laws are always keen to avoid. What is a trivial patent and when should a patent application be included in this category? Answering this question is very complicated. In practice, it entails the necessity for the proponents to compile a record of decisions on requests given by the patent offices, so as to facilitate the anticipation of what could be expected. Yet this does not exactly provide certainty. And uncertainty clearly works against the development of synbio, a biotechnology that depends on the investment of huge amounts of money. What will happen if uncertainty raises the number of lawsuits dramatically? Would the synbio market be able to survive in such a scenario?

3.3   Promoting open source

The most promising alternative to the patent system is the 'open source system', a framework mainly associated with informatics, which allowed, among other things, the successful development of Linux. The theoretical basis underlying the open source system is the construction of a product through its joint modification and improvement. Thus, open source applies to collaborative inventions which hold common property rights that are absolutely enforceable against third parties. In the field of biology, the open source idea was implemented, for instance, in the BIOS (Biological Innovation for Open Society) initiative founded by geneticist Richard Jefferson.

In the specific field of synthetic biology, MIT developed an initiative based on the philosophy of the open source system. A group of engineers and scientists from MIT, Harvard, and UCSF founded the BioBricks Foundation (BBF), which is "a 501(c)(3) public-benefit organization founded in 2006 by scientists and engineers who recognized that synthetic biology had the potential to produce big impacts on people and the planet and who wanted to ensure that this emerging field would serve the public interest." The ultimate goal of this organization is "to ensure that the engineering of biology is conducted in an open and ethical manner to benefit all people and the planet." (BBF 2015a). In order to achieve that aim, BIOS has developed three different projects. The first one, Stanford BIOFAB run by Drew Endy,

BBF founder and board president, "will map the central dogma (C-Dog) of yeast and contribute BioBrick™ parts to the public domain." The second, Global BIOFAB Network, is devoted to the creation of a network of BIOFABs to "create synergy and foster the development of community-driven technical standards and production of standardized biological parts." Finally, the SynBio Road Mapping Project, which is currently under construction, is defined as "an international effort of leading scientists, engineers, and practitioners in academia, industry, and government to present the grand challenges in synthetic biology." All these efforts are intended to "accelerate the development of standardized biological parts contributed to the public domain, foster the adoption of shared technical standards and ethical values, involving developing as well as developed nations" (BBF 2015b). From a legal perspective, it is important to highlight that all these initiatives have been made possible thanks to the design of a legal tool, the BioBrick™ Public Agreement, which includes a Contributor Agreement and a User Agreement. In this manner, those who join a BIOS "concordance" agree "not to assert IP rights against each other's use of the technology to do research, or to develop products either for profit or for public good. BiOS-compatible agreements can support both freedom to operate and freedom to cooperate" (Rutz 2010). In order to provide efficient support to the whole system, the MIT also created the iGEM Registry, which is "a growing collection of genetic parts that can be mixed and matched to build synthetic biology devices and systems" (iGEM, n.d.).

Thanks to all these efforts, the idea of 'open source' has gained significant influence in the field of synthetic biology. Currently, it is an intellectual option widely accepted by many of those working in this field. As noted by Henkel and Maurer (2007), "if open source initiatives can produce a particular part, then society will be better off without patents." The main issue is that, unfortunately, it seems inconceivable that open source will be able to fulfill the requirements of IP protection in an optimal way without any support from patents. Thus, it needs to be highlighted that the notion of the open source approach being incompatible with the patent system is a common fallacy. On the contrary, it will be much more effective if it is combined with a patent system. The most challenging task to be addressed by policy makers is, consequently, the design of an optimized legal framework to stimulate the development of this scenario. We really need to be creative in order to ensure that firms who are willing to cooperate in pools have their IP rights protected in some way. This task implies stable financing in the creation of the pools and the provision of legal advice to those willing to take care of them, but at the same time it calls for sufficient flexibility to change the legal framework whenever we encounter a situation that cannot be resolved with the current one. Sometimes it will be necessary to enforce anti-trust legislation, so as to prevent the constitution of a monopoly in the field of synbio. At times, it will make sense to use general legal principles in order to combat those who abuse the system, for example attempting to threaten participants with unjustified lawsuits.

4 A final remark

If we were to arrive at a final conclusion, it should be as follows: it is still too early to identify optimal solutions to all the questions relating to IP issues and synbio. It is still very difficult to determine whether synbio will increase in complexity in the same way as systems biology or whether, on the contrary, it will be much easier to handle than currently foreseen. It is also hard to foresee whether the synbio market will be built on the basis of cooperation between the participants or not. This key factor cannot be altered by changing the legal framework. However, it is perfectly possible to utilize our legal tools to stimulate cooperation and to avoid a general stalemate situation, which would be detrimental. To this end, it seems excessive to ask that the patenting system should be substituted by some other kind of regulation (cf. Miguel Beriain 2014). Instead, I believe it is essential to combine it wisely with these alternative tools, thereby creating a framework which protects IP rights when necessary and permits access to them whenever possible. It is not easy to balance both factors, but it is something we are obliged to do.

References

BBF (2015a). BioBricks Foundation. About. http://biobricks.org/about-foundation/. Accessed: 11 May 2015.
BBF (2015b). BioBricks Foundation. Technical Program. https://biobricks.org/programs/technical-program/. Accessed: 12 May 2015.
EC (1998). European Council. Directive 98/44/EC of the European Parliament and of the Council of 6 July 1998 on the legal protection of biotechnological inventions, OJ 1998 L 213/13.
Henkel, J., & Maurer, S.M. (2007). The economics of synthetic biology. *Molecular Systems Biology*, 3, 117. doi: 10.1038/msb4100161.
iGEM (n.d.). International Genetically Engineered Machine Foundation. Registry of Standard Biological Parts. http://parts.igem.org/Main_Page. Accessed: 11 May 2015.

The Manchester Manifesto (2009). *Who owns Science? The Manchester Manifesto*. Institute for Science, Ethics and Innovation at the University of Manchester. http://www.isei.manchester.ac.uk/TheManchesterManifesto.pdf. Accessed: 11 May 2015.

Miguel Beriain, I. d. (2014). Synthetic Biology and IP Rights: In Defence of the Patent System. In: I. de Miguel Beriain, C.M. Romeo Casabona (eds.), *Synbio and Human Health: A Challenge to the Current IP Framework?* (pp.201-209). Dordrecht: Springer Netherlands.

Patent troll (2015). In: Wikipedia. http://en.wikipedia.org/wiki/Patent_troll. Accessed: 11 May 2015.

Rai, A., & Boyle, J. (2007). Synthetic Biology: Caught between Property Rights, the Public Domain, and the Commons. *PLOS Biology*, 5(3), e58. doi: 10.1371/journal.pbio.0050058.

Rutz, B. (2009). Synthetic biology and patents. A European perspective. *EMBO Reports*, 10(S1), 14-17.

Rutz, B. (2010). Patent issues in SynBio applications. Presentation at the Synthetic Biology Workshop *From Science to Governance*, Brussels, 18-19 March 2010. http://ec.europa.eu/health/dialogue_collaboration/docs/ev_20100318_co18.pdf. Accessed: 11 May 2015.

Sapience (n.d.). *Applying for Patent*. Sapience – home of idea. http://www.sapience.org.uk/Services.aspx?Id=7. Accessed: 12 May 2015.

Universal Declaration of Human Rights (1948). UN General Assembly. http://www.un.org/en/documents/udhr/. Accessed: 12 May 2015.

Universal Declaration on the Human Genome and Human Rights (1997). UNESCO General Conference. http://unesdoc.unesco.org/images/0011/001102/110220e.pdf#page=47. Accessed: 12 May 2015.

III
Opportunities, Risks, Governance

Beyond unity: Nurturing diversity in synthetic biology and its publics

Sacha Loeve

"If the kids are united then we'll never be divided"
Sham 69, 1978

1 Introduction

Alarmist concerns have recently been voiced over the lack of unity of synthetic biology. In its May 2014 issue, the prestigious journal *Nature* featured a special section entitled "Beyond divisions: Building the future of synthetic biology" (*Nature* 2014). It is argued that, in the absence of commonly agreed definitions, standards of practice, and shared goals, the advancement of synthetic biology may be impeded and its promises harder to deliver, which may in turn deteriorate its public image and hamper its social acceptance. The epistemic, methodological, and cultural diversity of synthetic biology is even seen as a threat: the threat of "tribalism".

Should these issues be taken for granted? Much like other technosciences (for example, materials science or nanotechnology), synthetic biology is not a single scientific discipline, but rather an umbrella term gathering together a wide range of different techno-epistemic strategies arising from a variety of disciplines. Nowadays, epistemic diversity (scientific controversies, multi- and inter-disciplinarity) is commonly regarded as a positive factor for the advancement of research. The methodological diversity of technoscientific research is also positively considered in our so-called 'age of innovation', because it enables a larger range of possible innovation routes. As to cultural diversity — say, for instance, of diverse conceptions of life or diverse conceptions of the common good —, one may reasonably think that it constitutes a prerequisite for a 'democratic' societal debate. In view of these rather commonsensical arguments, such an urge to 'stand united' is rather surprising. Why does a unique identity matter? And why on earth would the establishment of a unified discipline allow for a 'better' synthetic biology in both epistemic and social terms?

In what follows, I start by taking stock of the astonishing diversity of synthetic biology. I then analyze the calls for unity as presented in the aforementioned special issue of *Nature*: as calls for a *disciplinary* unity of synthetic biology. I go on to contrast the seemingly disciplinary model of community-building advocated for synthetic biology with the overtly *multidisciplinary* model of community-building in nanotechnology, a field in which the majority of researchers are still firmly affiliated to their home disciplines and reject the prospect of a converging 'meta-discipline'. Based on the observation that the two technosciences display much comparable multidisciplinary dynamics, I argue that in the case of synthetic biology exhortations to unity are *not* driven by traditional disciplinary norms and ethos. They rather pertain to funding, industrialization, market normalization, and last not least, to a view of *the public as a threat*. I conclude with a plea to value and publicly expose the epistemic as well as cultural diversity of synthetic biology. Nurturing diversity in synthetic biology and its publics should better allow possible choices of society to be identified, debated, and decided upon.

2 Welcome to the Jungle

The first editorial of the *Nature* special issue (Figure 1), entitled "Tribal gathering",[1] deplores that, fourteen years after the publication of the field's seminal papers (Szostak et al. 2000; Elowitz and Leibler 2000; Collins et al. 2000), "the scientists who call themselves synthetic biologists already have disparate goals. By many accounts, 'synbio' is less a coherent discipline than a collection of tribes under the same name. They do not interact with each other nearly enough" (*Nature* editors 2014, p. 133).

---

1 The special issue features three editorials: the first, from *Nature* (*Nature* editors 2014), stresses the concern about "tribalism"; the second, from *Nature Reviews Microbiology*, emphasizes the firm grounding of synbio in microbiology (*Nature Reviews Microbiology* editors 2014); and the third, from *Nature Methods*, argues that realizing the promises of synthetic biology will require further efforts in establishing the basic foundations of the field and its standards of practice (*Nature Methods* editors 2014).

Beyond unity: Nurturing diversity in synthetic biology and its publics

Fig. 1 Cover of the Nature special issue "Beyond divisions: Building the future of synthetic biology," May 2014.

The editorial evokes the disparities between three sets of means and goals: "designing genetic circuits to make microbes do fun or useful tricks" (Levskaya et al. 2005), synthesizing biomolecules that incorporate non-natural building blocks for "revising the chemistry of life" (Benner 2004), and "stitch[ing] together stretches of DNA to create whole chromosomes" (Annaluru et al. 2014). Actually, these subdivisions (here between genetic engineering, xenobiology, and minimal cell creation) provide only a simplistic and quite US-centered picture of the field. At the least, synthetic biology gathers together approaches as diverse as metabolic engineering, minimal genome research, abiogenesis, bioinformatics, directed evolution, protocell chemistry, protein design, and biophysical methods. The scientists who engage in synthetic biology come from disciplines as diverse as virology, biophysics, astrobiology, synthetic chemistry, chemical engineering, or software engineering. The field of synthetic biology is far more diversified and in motion than suggested by the now usual distinctions between a handful of main sub-fields such as DNA-based device construction, genome-driven cell engineering, protocell creation (O'Malley et al. 2008), or by the schematic dichotomies between top-down and bottom-up synthetic biology.

A recent workshop co-organized by Paris Sorbonne University and Freiburg University, Worldviews and Values in Synthetic Biology,[2] revealed a great diversity of ways of doing synthetic biology. The organizers conjectured that this might be because mainly European approaches were represented, and that the European profile of synthetic biology proves far more diverse and multifaceted than the dominant US vision of the field. The techno-epistemic strategies discussed in the workshop included: combining genetic modification and imaging to allow real-time control of cells by computer (Batt 2014); enabling artificial protocells and natural cells to "talk with each other" (Mansy 2014); building cellular 'mimic' organelles to study the interaction of natural cells with pathogens for medical purposes (Römer 2014); hybridizing synthetic chemistry, nanotechnology, and biochemistry; addressing the challenge of synthetic regulation (Pompon 2014); designing programmable evolution machines to accelerate the artificially-evolved production of de novo biomolecules (Jaramillo 2014); devising an unprecedented toolbox for chemists (Jullien 2014); and, not to forget, the biosynthesis of tons of profitable molecules in industrial production plants (Delcourt 2014) — a trend of industrial synthetic biology where the supposedly 'emerging' technology appears to be an extension of the already well-established sector of industrial biotechnology.[3]

The workshop also questioned the popular vision that synthetic biology is in essence 'the engineering of biology'. In practice, the nature of the relationship between engineering and biology in synthetic biology is a subject of ongoing debate (Frow and Calvert 2013). For these disciplines "are typically seen to involve different ways of knowing and doing, and to embody different assumptions and objectives" (ibid., p.42). A similar tension can be found between engineering and synthesis as two different ways of learning by doing (Benner et al. 2011; Luisi 2011; Malaterre 2013; Loeve 2014).[4] Should these tensions be despised as symptoms of "tribal divides",

---

2 See http://synenergene.eu/resource/summary-report-paris-workshop-worldviews-and-values-synthetic-biology.

3 For an already well-furnished list of synthetic biology companies see Schmidt (2012, pp.227-230).

4 Pier Luigi Luisi makes an epistemological distinction between the chemical, synthesis-centered trends of synthetic biology and "biological engineering," which is "clearly and purposely directed towards one goal set from the start" (Luisi 2011, p.343). In a similar fashion, Steven Benner distinguishes between (chemical) synthetic biologists using unnatural molecules to do things that natural biology does, and genetic engineers assembling natural bio-parts to do things that natural biology does not (Benner et al. 2011, p.69). Biological engineering rests on the ideal of beholding a fully predictive

as suggested by the above-mentioned editorial of *Nature* (*Nature* editors 2014)? In practice, they do not impede research. Rather, they open the way to a diversity of challenging combinations between components that do not straightforwardly fit together.[5] The tensions between engineering and biology, evolution and design, or engineering and synthesis are not sterile divides, but *creative tensions*. In bioinformatics, for instance, computer scientists are interested in 'formalizing' biological behaviors precisely because they (or at least some of them) see the biological as a medium recalcitrant to formalization, and *not* because they believe in a 'natural' analogy between the cell and the computer.[6] Contrary to a computer, which strictly follows formal rules, there is no a priori formalizer in the biological medium (Delaplace 2014). In other terms, bioinformaticians are interested in 'programming life' precisely because life is not 'preprogrammed to be programmed'. This is a challenge that demands revisiting the design of computer language.

To diversify even more this journey into the 'synbio jungle', regional differences in research agendas and cultures need to be emphasized (Engelhard 2014). A number of German researchers are rather reluctant to adopt the label 'synthetic biology' (Römer 2014), either because they have found it to have bad connotations, or because their biosynthetic constructs are not fundamentally centered on genetic engineering, or both. Others stress that synthetic biology can do without genetic engineering (Lentini et al. 2014). In Italy, where the chemical tradition of synthetic biology is very strong, synthetic biologists have little contact with industry.[7] While some approaches of synthetic biology are rather aggressively market-oriented (a

---

model *before* synthesizing the system: synthesis is only the assembly of well-specified, predictable, and functionally unambiguous parts; it is the application of a model. As noted by Bernadette Bensaude Vincent (2013, pp.126-127), synthesis, in this case, means nothing more than reverse analysis. On the contrary, synthesis in chemical synthetic biology is not only the reverse side of analysis (Benner and Sismour 2005): it is a heuristic method allowing the gaining of new knowledge that is neither analytical nor theoretical knowledge, but functional, operative knowledge, or simply "synthetic knowledge," as Christophe Malaterre (2013) put it. According to the latter, "synthetic knowledge" is not restricted to "knowledge-how" (knowledge of the means suitable for a certain end: "how to replace some components of DNA"). It also delivers "knowledge-why": knowledge of the relationships linking properties and functions of a living system's components in a causal model—not a knowledge of means-to-an-end, but a knowledge of *effects*.

5 This suggests an extension of the meaning of "kludging" in synthetic biology (O'Malley 2009): not only kludging together unfitting biological parts, but also the kludging of unfitting epistemological components.

6 This, by the way, makes current bioinformatics rather different from traditional cybernetics.

7 Sheref Mansy, personal communication.

stance epitomized by the mediatic US figures of Craig Venter or Drew Endy), others remain very fundamental. In the latter case, synthetic biology is regarded either as a method for illuminating the 'universal' properties of life and the search for its origin beyond the 'provincial' manifestations of extant life (Attwater and Hollinger 2014), or as a method to understand why life is "like this and not like that" (Luisi 2014).

These more 'fundamental' approaches — those considering that synthetic biology has something to teach us about life (and which, paradoxically, are often *chemical* approaches) — are more common in Europe than in the US, where bioengineering reigns supreme. Yet bioengineering, with its characteristic focus on reproducible processes and on the decoupling of design from fabrication, of parts from each other, and of synthetic bacteria from their environment, falls under the category referred to by the philosopher Hugh Lacey as the "decontextualized approach" of technoscience (Lacey 2012). By contrast, chemical synthetic biology appears to be more context-sensitive, for it must pay attention to the peculiarities of the environments that living processes require and also create while emerging, maintaining themselves, and setting their own purposes (Luisi 2006). To some proponents of the chemical approach, the various environments of interest extend even to the *cultural and historical contexts* in which we interact both cognitively and practically with living systems.[8] The chemical approach is thereby more prone to acknowledge the culturally differentiated and embedded character of synthetic biology than bioengineering, which aims at 'making (living) things the same.'[9]

This glimpse into the jungle of synthetic biology reveals not only a diversity of means and goals, but also and more fundamentally a diversity *of aspirations* — of different meanings granted to 'synbio' research. Unfortunately, the diversity is silenced in public discourses.

---

8   According to Luisi, this is true of the definition of life, of which the "changes automatically reflect how science values have moved with time." Today, "the different definitions of life reflect the main schools of thought that presently dominate the field on the origin of life" (Luisi 1997, p.613). The notion of *purpose* in biology is also context-dependant: "The notion of purpose is not objective, but it is contextual, changing in time and is probably different in different societies and traditions and point of time" (Luisi 1998).

9   For instance, a recent viewpoint article on future perspectives for synthetic biology (Church et al. 2014), gathering together only US geneticists (George Church) and biological engineers (Michael Elowitz, Christina Smolke, Christopher Voigt, and Ron Weiss), does not even mention the existence of several different approaches to synthetic biology, whereas chemists like Luisi or Benner always mention alternative approaches.

3 Disciplining Synthetic Biology

"Beyond divisions: Building the future of synthetic biology" (*Nature* 2014): the title reads like a slogan. In its email advertisement, Nature Publishing Group insists on the deliberately proactive aim of the special issue. For it not only "charts the progress of a nascent multidisciplinary field," but also explores its "challenges in developing clear goals, standards of practice and pathways to lucrative commercialization" (*Nature* Publishing Group, 2014). Of course, this is not the first time that a high-impact journal such as *Nature* or *Science* has taken the lead in actively promoting an emerging technoscientific community and helping to articulate its visions and agendas — this was the case for nanotechnology,[10] for graphene,[11] or for optogenetics.[12] It would be at least naïve to believe that the role of scientific journals is solely limited to *reporting* state-of-the-art research results, and is not to *support* emerging promising fields at an early stage. Nowadays, scientific journals definitely shape the successive waves of technoscientific fashions (Potthast 2009). However, that is not so much the point here. Far more interesting is the *way* by which the journal chose to do so for synthetic biology, namely by expressing a strong concern over its lack of *unity as a discipline*.

Indeed, far from being regarded as creative tensions, the differences between synthetic biologists are looked upon as unfortunate obstacles to overcome:

> Divisions run deep. A synthetic biologist may come to a project as an iterative tinkerer, a methodical engineer or an intuitive explorer. Engineering types can be flummoxed by unpredictability, and sciencey types frustrated by demands for inflexible definitions and standards. Some researchers have focused on establishing standards so that various "gear parts" can be easily interchanged and combined. Others resist, arguing that variability among cells and circuits defies human-devised specifications. Some scientists push for open access, the better to mix and match inventions; others call for strong protections for intellectual property, the better to incentivize development

---

10 Explicitly with the foundation of the journal *Nature Nanotechnology* in 2006.
11 After having twice rejected the seminal graphene paper by Novoselov et al. (2004), *Nature* reconsidered its initial rejection and published a paper by Geim et al. as early as in 2005 (Novoselov et al. 2004). *Nature* then became proactive in spreading expectations about "the graphene 'gold rush'" by publishing papers with titles as eloquent as "The rise of graphene" (Geim & Novoselov 2007) or "The quest for supercarbon" (Peplow 2013). After the Nobel Prize for graphene was awarded to Geim and Novoselov (2010), the journal even published the first "roadmap for graphene" (Novoselov et al. 2012) which is now the spearhead of the Graphene Flagship project funded by the European Commission as part of its "Horizon 2020" Projects.
12 Promoted as "method of the year 2010" in the ten years anniversary issue of *Nature Methods* (Pastrana 2011).

of useful applications. The paucity of connection between these groups has provided ample fodder for sociologists of science. [...] Whatever the reason, the discipline suffers from its divisions (Nature editors 2014, p.133).

Most of the special issue is then devoted to the search for potential remedies to these "tribal" pathologies. In particular, a comment section entitled "How to best build a cell" gathers advice from "experts" who recommend solutions to transcend such divisions.

The first comment, "Bring in the biologists", stresses that one must first know more about biology in order to make synthetic biology "a predictable engineering discipline" (Colins 2014, p.155). Although "biologists and engineers should learn to learn from each other [...], overcome cultural differences and biases [and] move beyond dismissive, tribal comments such as, 'Well, she's not a biologist' or 'That's not engineering,'" the dissymmetry is nevertheless patent: "biologists need not feel compelled to 'do engineering.'" The comment expresses a need for "disciplining" synbio by bringing it closer to a really biology-minded and less engineering-inspired discipline. Ironically, another comment provides just the opposite advice: "Automate efficient design" (Weiss 2014), which basically means "bring in the engineers" — more standardization and more black-box design tools to discipline the intrinsic variability and fuzziness of biology. Another expert recommends a research direction that would break up with the mainstream microbe-oriented character of synthetic biology: "Make tools for mammalian cells." "For synthetic biology to be useful in medicine, more and better tools are needed for work with mammalian cells. The tools that are now standard for bacteria are missing or underdeveloped for mammalian cells. [...] No matter how sophisticated the tools, bacteria will not be up to this task" (Fussenegger 2014, p.157). This recommendation stands in stark contrast with the second editorial of the special issue, which puts forward the firm grounding of synthetic biology in microbiology (*Nature Reviews Microbiology* editors 2014). Another expert's advice, "Capitalize in evolution", stresses that "engineers may have little time for evolutionary theory" while "evolutionists may have little interest in engineering". However, "scientists must consider both together to get the most from synthetic biology" (Sauro 2014, p.157). Another's recommendation is to "Build green 'bioalchemists'"(Ellington 2014), that is, an armada of engineered bug "brokers" that would metabolize various feedstock to turn them into a wide range of valuable goods,[13] thereby contributing to aligning synbio with the widely accepted goal of "greening" the economy.

---

13 These bugs would be "carbon workers" in the sense in which Jennifer Gabrys puts it to refer to the crowd of nonhuman beings processing the biodegradation of plastics (Gabrys 2013). As she recalls, the term "carbon workers" is commonly used across climate change

The irony of all these recommendations is that in order to spur unity they all put forward divergent strategies based on inescapable distinctions: more biology for some *versus* more engineering for others, more evolution *versus* more preprogrammed design, more mammalian tools *versus* more bacterial tools, etc. Against this backdrop, one might doubt that 'synbio' will ever be able to recover from its allegedly 'pathological' divisions. This is not surprising, since the experts consulted are all distinguished academic professors of microbiology, biochemistry or bioengineering, each one defending their respective research strategies — with the exception of one science policy strategic advisor,[14] whose advice is on the contrary to "Agree on a definition" (Maxon 2014). Is synthetic biology a new discipline of engineering or an extension of biotechnology? Has Craig Venter created artificial life or just a sophisticated genetic tool? "These ambiguities complicate discussions among scientists, hinder policy-makers, impede efforts to fund synthetic-biology research and thwart regulation that might build public confidence. […] Without a consensus on what synthetic biology is and is not, informed policy is difficult to set." (ibid., p.156) If only life could be less complicated and uncertain or, in a nutshell, 'easier to engineer'!

To what extent is the observation of synthetic biology as a 'tribal gathering' justified? The observation does indeed seem justified to some extent. According to a recent scientometric inquiry conducted by Benjamin Raimbault and Pierre-Benoît Joly (Raimbault et al. 2013), the quantitative data concerning institutional dynamics and researchers' trajectories gathered together under the umbrella term of synthetic biology indicate an increase in population, productivity, and connectivity around 2008, followed by a stabilization phase during which distinctive groups were formed along the lines of identifiable techno-epistemic clusters such as "minimal genome", "protocell synthesis", "device-oriented approaches", "compound-oriented approaches", "computer/cell approaches", "cell-cell interactions", etc. The result, therefore, is a rather 'tribal' landscape structured in groups having strong epistemic identities. However, the groups formed suggest that this tribal organization is in fact partly due to a few 'star' researchers who bring resource and credibility to synthetic biology. Indeed, eight authors account for more than 60% of the main articles cited since 2004! George Church, Craig Venter, Drew Endy, Jay Keasling, Rob Carlson, and other prominent figures of 'synbio' play a key role as 'institutional entrepreneurs' beyond the scientific community and engage in

---

literature to refer to the "services" carried out by the "carbon sinks" that are designated by the Kyoto Protocol.
14 Mary Maxon, Director of Biosciences Strategic Planning and Development, at Berkeley National Laboratory (Maxon 2014).

activities related to innovation policy, intellectual property, public relations, and ELSI (Ethical, Legal and Societal Implications). Accordingly, the problem is not so much one of 'tribalism' per se; it is rather one of 'starification' (Figure 2).

Fig. 2   Snapshots from the video "Synthetic Biology Explained".[15]

In addition to the phenomenon of 'starification', the problem behind 'tribalism' is also perhaps 'gamification', in the sense that the iGEM competition[16] is first of all *a game*, and because this game plays a structuring role of breeding a typical synthetic biologist's ethos (young, cool, dabbler, undisciplined, adventurous, competitive, enterprising...) without aiming to build a definite and sustainable sense of community so far. As Sara Angeli Aguiton (who followed a French team during an iGEM competition season) argues, the students who invest themselves heavily in an iGEM project for six months do not necessarily come to swell the ranks of the 'synbio' community. They train themselves to a certain practice and research style, raise funds, complete the project as quickly as possible, create a Wiki, feed the repository of DNA bio-parts, participate in the jamboree, obtain an award... and then most of them simply go back to their home community to complete their education (Angeli Aguiton 2010). They play a game, enter in a space apart where new rules prevail; they literally *perform* synthetic biology for a limited amount of time. When the performance is over, the kids come back home.

One might be tempted to conclude that synthetic biology presents itself with two faces: on the one hand, it displays the wild and undisciplined character of a diverse 'jungle'; on the other, 'synbio' dreams of fusing its disparate elements into a single 'organic' community standing as one discipline. Its dual profile somehow reflects the ambivalent relationship synthetic biology entertains with life: on the one

---

15 "Synthetic biology: Putting the engineering back into genetic engineering", written, animated, and directed by James Hutson, produced by Bridge8, and funded by the Australian Government (https://www.youtube.com/watch?v=rD5uNAMbDaQ).
16 The "International Genetically Engineered Machine Competition".

hand, praising life's prolixity, mutability, and diversity as a rich source of potential technological solutions; on the other, seeking to simplify, to standardize, and to discipline life's operating principles to make it suitable for engineering. That is why synthetic biology is often named a 'discipline' in a rather incantatory manner, as a 'nascent discipline', a 'maturing discipline', an 'emerging discipline', *invoked* as a new discipline of the future, not *described* as a new discipline here and now.

However, such an effort to foster a disciplinary identity despite the multidisciplinary profile and epistemic pluralism of synthetic biology is not a unanimous trend of the field. For the most part, it reflects the values of the small group of American bioengineering activists gathered around Drew Endy, Jay Keasling, and Rob Carlson, who have promoted the field as a "new professional engineering discipline" at the almost annual Synthetic Biology X.0 conferences since 2004 (Campos 2009; Bensaude Vincent 2013a) with an obvious willingness to make a clean sweep of the past.

4 Multidisciplinary Community-Building: Nanotechnology

In stark contrast with the *Nature* special issue "Beyond divisions", the editorial of the first issue of *Nature Nanotechnology*, in 2006, was entitled "Small is different" (*Nature Nanotechnology* editors 2006a). The text was a "teaser", presenting nanoscale science and technology as an inclusive field offering "opportunities for all sorts of scientists and engineers." Instead of deploying a grand narrative celebrating the founding fathers and the milestones of nanotech, the editors overtly stated that "nano" has no unique identity, definition or historical origin: "Depending on who you ask, nanotechnology started in 1981, 1974, 1959 or the Bronze Age. And depending on who you believe, and the definitions they use, the world market for nanotechnology products will be worth $2,600 billion in 2014, or $1,000 billion in 2015." The editorial's main rationale was that "nano" is attractive not in spite of its lack of unity, but precisely *because* of it. In other terms, there is plenty of room for a diversity of researches at the nanoscale. The only consensus on the meaning of nanoresearch is that "nano" is different, in many respects. First, the sole criterion for "something — a material, device, system or process — to be truly nano," is that "the relevant length scale must be small enough for its properties and behaviour to be different from those observed in the bulk." Second, "nanoscience and nanotechnology are genuinely multidisciplinary and broad-based subjects." Thus "Small is different" does not only mean that "nano" is different from bulk, but also that "nano" is different from "nano". Third, "many researchers have rebranded what they have been doing for years as nano" in order to take advantage of the "nanolargesse" of governments and companies. There is nothing new or wrong in that, the editors confess, it is just a means for researchers to continue doing "basic scientific research under the banner of nanotechnology" while getting "new money for building new laboratories and facilities to explore material and processes on the smallest length scales." Thus, "nano" is different... even when it is the same!

Fig. 3   Cover of the first issue of Nature Nanotechnology, October 2006.[17]

The first issue of *Nature Nanotechnology* also featured another editorial piece gathering together thirteen interviews of researchers expressing thirteen different

---

17   The main themes listed were selected to be as diverse as possible: "NanoSQUIDs" (Superconducting QUantum Interference Devices, i.e., very sensitive magnetometers); "Carbon nanotubes—Sorted"; "Bionanoelectronics—Viruses as switches"; "Molecular motors—Work in progress."

opinions on what nanotechnology means to them (*Nature Nanotechnology* editors 2006b)! The editors gave it a weird-looking title, "nan'o·tech·nol'o·gy *n*.," whose calligraphic form was intended to reflect the rhapsodic diversity of the field itself. The collection of interviews exhibited not only different definitions of nanotechnology but also different *ways* of approaching its definition, either by the tools or by the finalities, by insisting either on continuities or on ruptures, on broad aspects (e.g., burgeoning capabilities of building things at the nanoscale) or on unique characteristics (e.g., imagining and manipulating nanoscale objects one by one), sometimes by integrating the political, economical, and social aspects of the 'nanoworld'. Remarkably, instead of demarcating and welding a community, the journal chose to immediately praise the disunity, the lack of neutrality, and the interested nature of any definition of the field.

While nanotechnology promotes interdisciplinarity, most of its researchers were (and still are) reluctant toward the prospect of a converging 'meta'-discipline. Such a prospect has so far been formulated once in the framework of the NBIC program launched by the US National Science Foundation and Department of Commerce (Roco and Bainbridge 2002).[18] The NBIC report announced nothing less than a renaissance of a unifying science at the molecular scale that would lead to a "technological convergence". All the NBIC-related disciplines were impelled to combine their elementary constituents (bits, atoms, neurons, and genes: the 'small b.a.n.g.') while working in a synergetic manner to enable the generation of new tools for technologically "enhancing human performance" at a physiological, spiritual, and social level. The authors of the NBIC report dubbed the technological convergence a "new renaissance" because it was supposed to bring about a holistic synergy of all fields of science and engineering at the molecular level.

Actually, most nanoscale researchers (at least the majority of them in France) have simply *ignored*—or pretended to ignore—the converging technologies program. When asked about it, most of those who knew about the NBIC report declared that they were opposed to its vision, not only because of its obvious transhumanist inspiration, but also and above all because of its view of technological convergence as a meta-disciplinary framework putting an end to disciplinary divisions. As a measure against the project of convergence, some research groups even advocated a rebel strategy of 'divergence'.[19]

---

18 NBIC stands for nano-bio-info-cogno.
19 Such as the NanoBioSystems group in Toulouse, whose research is rather close to that advocated in the name of 'NBIC convergence' (they develop generic nanotech tools that bring 'bio' and 'info' increasingly closer to each other) and perhaps precisely *because* of this proximity.

Has the persistence of a robust (multi)disciplinary matrix ever represented a threat of 'tribalism' for the nanofield? Did it lead to the atomization of the field?

In their study of the epistemological dynamics of nanoscale research, Anne Marcovich and Terry Shinn (2014) argue that nanotechnology can be described neither in terms of over-differentiation of domains nor in terms of a 'seamless' inter- or trans-disciplinary regime such as the one advocated in the name of "mode 2 research" (Gibbons et al. 1994). Marcovich and Shinn explore a third way that they call "new disciplinarity". The new disciplinarity is not a strictly traditional disciplinary regime, but it is not a post-disciplinary one either. It is a regime of research where disciplinary boundaries are firmly maintained while still having elastic and moving demarcations. Its practitioners have both strong and flexible bonds with their home discipline; they maintain an "elastic" connection with it. According to Marcovich and Shinn, the maintenance of disciplinary boundaries is a pledge of plurality and dynamism for the nanofield. The disciplines are sufficiently robust to allow practitioners to cross their boundaries for short periods, during projects mobilizing practitioners from other disciplines and addressing questions that cannot be resolved in a single domain. However, the *way* in which each practitioner engages in such projects remains rooted in the question that each one asks and pursues from the standpoint and in the grammar of their home discipline. In so doing, "scientists speak across the borderland whose structures allow them to remain inside their home discipline" (Marcovich and Shinn 2014, p.12). The disciplinary/interdisciplinary tension thus resolves itself into a multiplicity of temporalities: short-term temporality for project work, long-term temporality for the lives of disciplines; sometimes, when a project assumes a synergistic character, an intermediate temporal regime develops, and the disciplinary question evolves.

Moreover, the new disciplinary regime according to Shinn and Marcovich grants special importance to *combinatorials*. Combinatorials of instruments, materials, concepts, and people play a structuring role in the cognitive trajectories of researchers. Such combinatorials are temporally expressed in a pulsatile phenomenon that Marcovich and Shinn call "respiration": it is "a pause, a breath during which scientists evaluate, take stock of what new combinatorials are available for development of their present research project, or alternatively, that can be mobilized for initiation of a new project" (ibid., p.11). "Respiration" plays out in two spheres: "concentration", where researchers funnel new combinatorials towards the accomplishment of a particular techno-epistemic research aim (for instance, when a particular research question involves recruiting new materials, instruments, and people in order to be addressed); and "extension", where researchers address new fields and inscribe their research in wider horizons (for instance, when a physicist ventures into biological subject-matters). The circulation of practitioners within and between spheres of concentration and extension is punctuated by these crucial intervals of respiration—a vivid and structuring dynamic which is possible not despite, but *because* of the coexistence of a multiplicity of differentiated disciplinary approaches and skills.

To sum up, nanotech provides a compelling example of a techno-epistemic field that is thriving and flourishing without ever aspiring to unity. Far from threatening the field with explosion, the multifaceted and pluralistic aspect of nanotechnological research provides it with a consistency and a temporal allure of its own. It even allows it to breathe!

For sure, synthetic biology is a combinatorial field, too. It builds upon foundational research components borrowed from a wide range of preexisting fields, and it bases its engineering vision on the assumption that biological systems are inherently modular (Morange 2009). Without claiming that this framework would apply identically to synthetic biology, it is nevertheless clear that synthetic biology displays a much comparable multidisciplinary profile. It is thus all the more surprising that synthetic biology is often invoked as a new discipline — while nanotechnology is not. Indeed, the tensions between engineering and biology, design and evolution, or engineering and synthesis, makes the prospect of synthetic biology as a unified meta-discipline hardly plausible. Nevertheless, this does not mean that synthetic biology is a mere collection of tribes who do not talk with each other. Through participation in synthetic biology projects and teams, incoming scientists belonging either to a disciplinary or to a transitory research regime[20] interact with a field having leaky and moving boundaries. The interactions between individual scientists and the loosely defined field of synthetic biology frame individual research agendas as well as the moving contours of the field itself. Such diversity constitutes in no way a threat to the vitality of the field. It is part of its dynamics, and it is not at all obvious that it must lead to an excessive division—to a fragmentation—of the field.

At this point, it should be recalled that the choice of the label 'synthetic biology' (Campos 2009; Bensaude Vincent 2013a) is itself the result of a compromise between several trends constitutive of the field. The first proposed label was 'constructive biology' (as opposed to descriptive biology), but it was too narrowly associated with the community of artificial intelligence and bio-inspired robotics. The next label to be proposed, especially by Carlson, was 'intentional biology', but many biologists felt offended by it because it seemed to imply that what they had been doing so far was 'unintentional biology' (Campos 2009, p.18). 'Synthetic biology' was suggested

---

20 In the transitory regime of research according to Marcovich and Shinn (2012), scientists circulate back and forth between a disciplinary framework and an entrepreneurial environment, and often migrate back to the disciplinary referent.

by Carlo Bustamante to Endy and Carlson as an explicit analogue of 'synthetic chemistry' during a *Nature* cocktail party in 2001 in San Francisco. However, Bustamante's perspective stood very far from the central business of bioengineering, as he is a biological chemist and biophysicist engaged in single-molecule manipulation and visualization, a work typical of nanobiotechnology. Moreover, Carlson and Endy knew that the term 'synthetic biology' was not new. It had already been used by Steven Benner, a chemist, who was himself drawing on earlier uses of the term (Hobom 1980; Rawls 2000). Thus, Endy preferred to use the term 'engineering biology' (Endy 2005). However, the thought that biology could experience the same fate as chemistry when it shifted from analysis to synthesis was definitely a powerful one. So 'synthetic biology' finally met with general acceptance during the SB 1.0 Conference held at the Massachusetts Institute of Technology (MIT) in 2004. It was recognized as the best compromise between the militant 'engineering-first' approach typical of bioengineering and the MIT, and the huge diversity of biologists, biochemists, synthetic chemists, and biophysicists who were about to join the field. Accordingly, the label 'synthetic biology' results more from a concession to epistemic pluralism than from a willingness toward disciplinary unification.

5 Non-Disciplinary Drivers of Synbio's Unification

Ultimately, the plea to unite synthetic biology might have nothing to do with disciplinary norms and ethos. The attachment to scientific disciplines might well be vivid (as in nanotechnology), but it is an attachment to pre-existing disciplines, not to synthetic biology as a new discipline. The exhortation of synthetic biology to 'stand united' does not seem to answer to any alleged need to comply with the traditional ethos of scientific disciplines. If the drivers for uniting synthetic biology are not disciplinary drivers, what are these drivers, then?

These motives are not hidden structures that would have to be unearthed by some perspicuous sociologists of science. The proponents of synthetic biology as a so-called 'discipline' overtly set them out, as is visible in the *Nature* special issue "Beyond divisions".

5.1 Winning the Competition for Funding

Funding is one of the issues mentioned in "Beyond divisions". If synthetic biology does not appear to be something really new and unique, then it will not be supported by a tsunami of investments as in nanotech, brain science or bioeconomy, which are all supported by large national, federal or European initiatives.

> Emerging technologies that have more precise definitions have captured significant federal support, in terms of funding and policies. For example, the Networking and Information Technology Research and Development programme, created by Congress in 1991, is slated to receive funds of about US$3.8 billion in 2015, while the National Nanotechnology Initiative, created in 2001, could receive $1.5 billion (Maxon 2014, p.156).

As we have seen, this statement is rather inaccurate for nanotechnology: The field still has no precise definition and *does not need one*, at least for funding purposes.[21] Actually, the situation of synthetic biology is different: Its priority is to distinguish itself from all current molecular biology and biotechnology, hence the repeated claim that classical genetic engineering was actually mere empirical tinkering or handicraft (Marlière 2009). "If synthetic biology is the same as biotechnology, it is supported by other programs, and so does not need a large-scale, federally coordinated effort" (Maxon 2014, p.156). However, if synthetic biologists cannot agree on a definition, they are advised to agree on a *roadmap*, thereby following the example of the UK (UK Synthetic Biology Roadmap Coordination Group 2012), which resulted in the deployment of a large public funding framework one year later.

Synthetic biology constructs its novelty by promising the emergence of something more than the sum of the parts of twentieth-century molecular biology and bioengineering. Thus, synthetic biology is not just making new promises (when it does make promises, they are always the same, and not very different from those of nanotechnology); synthetic biology is addressing foremost the practical realization

---

21 For *regulatory purposes*, it is commonly agreed that regulatory bodies need to work with some definition, because for a toxicological assessment it is important to specify the relevant size range in which these objects are able to cross several biological barriers, as well as which of their characteristics contribute to cause nano-specific toxicological effects (Lacour 2011). For funding purposes, however, quite the opposite is the case: the definition is kept as inclusive as possible in order to attract a maximum of research areas in the nano field.

of the promise of *biology as technology*[22] (Carlson 2011; Mackenzie 2013a). Its promise is to enable the systematic engineering of biology at an industrial-system level.

5.2 Acquiring Capabilities for Successful Industrialization

The unity of the multidisciplinary field of synthetic biology is voiced not only in terms of a new biological academic discipline but also in terms of 'professional engineering'. Yet industrial-level engineering requires standardization as well as critical mass.

While synthetic biology is a combinatorial field, it relies on combinatorials only to the extent that it seeks to build something new, something more than the sum of its components, which Mark Bedau qualifies as "engineering desirable emergent properties" (Bedau 2014). For this reason, synthetic biology must not only focus on the components (the 'parts') but also pay attention to the *connections* between the components and to the rules that govern these connections. However diverse synthetic biology might be, it must agree on unified, generic standards in order to become real engineering and to reach an industrial-level critical mass. In this respect, the kind of novelty that synthetic biology is trying to negotiate is less epistemic than it is logistic, as Adrian Mackenzie argues in the case of biofuels (Mackenzie 2013b). Because industrial systems depend on networks in which interchangeable elements can constantly circulate and be exchanged, no industry is possible without standards. This is true, for instance, of nineteenth-century industrialization. Contrary to a popular vision, the industrial revolution was not due to the considerable power unleashed by the steam engine alone. It was due to the synergy between these machines and the setting up of a standardized production system of identically-shaped, accurately-sized, and interchangeable machines parts.[23] The steam engine, which had hitherto been confined to being used for the dewatering of mines, could be made movable because the standardization of parts allowed it to carry its technical associated milieu with itself. This is also true of the Internet. The

---

22 Drawing on the case of Drew Endy's reconstruction of the T7 bacteriophage (Chan et al. 2005), Allain Pottage remarks that biological parts are not only engineered once for a specific purpose, they are rather engineered to be "engineerable" (Pottage 2009, p. 171). The engineered T7 bacteriophage is not only a reconstructed bacteriophage, it is also a bacteriophage re-engineered to serve as an "engineerable" device outside the lab.
23 A synergy epitomized by the partnership between the engineer James Watt and the manufacturer Matthew Boulton, who later applied these techniques to the minting of coins, striking millions of identical (and harder to counterfeit!) pieces for Britain and other countries.

'network of networks' is not only a network of machines. It is also, and indissolubly, a multilayered architecture of standardized communication protocols[24] allowing various machines, devices, and software to talk with each other.

Without standardization and interchangeability, the engineering approach touted by synthetic biology will never stand because it will lack general rules to make the parts work together and to specify which parts should be used for which function. Biobricks, for instance, would remain conventional molecular biology. Hence one hears the repeated calls for unifying standards in order to make biology real engineering.

5.3   Structuring the Market

The divergence between those who seek to protect the fruit of their research by patents and the supporters of free-for-all tools in open access is referred to as a "cultural divide" in the *Nature* special issue "Beyond divisions" (Nelson 2014). For the advocates of intellectual property, patents are imperative to spur innovation and allow lucrative commercialization; for the advocates of open source, patents hamper innovation, and open access is the way forward to allow all bio-parts makers to converge on shared and effective standards, as has occurred with the Internet. This divide supposedly reflects the tensions between the respective intellectual-property (IP) cultures of biotechnology (with its tradition of patenting) and of software engineering (with its tradition of open sourcing and sharing). Moreover, ethical considerations add to the "cultural" nature of the controversy: is it ethical to patent every single parcel or product of nature? (The general reply is that synthetic biology creates DNA that does not occur naturally.) If something goes wrong, who is responsible for the malfunctioning design of an open-source bio-part? (The general reply is that a system of free copyright could well do the job).

In order to bridge the gap between these two IP cultures, one way forward is often suggested: "share and protect" (Nelson 2014, p.154). It consists of making bio-parts freely available, as 'commons', whereas built-up devices made from those parts would be eligible for patenting. The use of the 'Lego parts' would be free; only the complexity of the assembly and design efforts would be rewarded. This "share [the parts] and protect [the design]" vision assumes that complete decoupling of fabrication and conception is realizable or, in other terms, that the materiality

---

24   HTTP (HyperText Transfer Protocol), HTML (Hypertext Markup Language), TCP-IP (Transmission Control Protocol/Internet Protocol), etc.

of the biological material could be rendered perfectly transparent to the human (computer-aided) mind.

Is it really a debate between 'two cultures'? As Jane Calvert remarks, the partisans of both open source and proprietary regime make similar use of informational and computational metaphors to define the entities whose suitability for appropriation is to be discussed. Most of them agree on a 'diverse ecology' to make place for both regimes (Calvert 2012). It is also anticipated that future synbio products are likely to be 'complex technologies', much like electronic hardware, where IP rights are hard to identify,[25] with distributed and fragmented ownership, and sometimes overly broad. In the microprocessor or mobile phone industry, no manufacturer owns all the patents that cover its products (Henkel and Maurer 2009). The complexity of IP rights often pushes industrial companies to share their technology instead of using patents to exclude competitors, as occurs in the pharmaceutical industry. They either negotiate (in financial terms) cross-licensing agreements or, when the transaction cost is too high, they may choose to share their patents on a collaborative 'give-some-get-some' basis, while protecting the patents covering their designs and production processes (or while keeping them secret).

The debate between the two IP regimes should not be taken to be something it is not. Despite the sincere enthusiasm of young iGEM students or garage biologists inspired by the hacker culture, the open-source regime is not a radical alternative to capitalism. The small synthetic biology start-ups confronted with powerful multinationals that dominate the biotechnology market require venture capital, and they need to file patents to justify their investments. In this highly capitalistic landscape, the inclusion of an open-source layer in the IP regime is not just a matter of reclaiming an ideal of the common good. It is also a convenient way to boost innovation by dropping the production costs of bio-devices: a true neo-Fordist approach. Accordingly, attempts to settle the controversy between the open versus the proprietary regime aim not only at surmounting a 'cultural divide'. They traduce the search for coherent market norms that would secure an optimal trade-off between openness and profitability, in a market-based approach to the unification of the field.

---

25 It is actually already the case with the iGEM's Registry of Standard Biological Parts, which contains 'open-source' bio-parts embedding DNA sequences already covered by patent claims (e.g., the sequence coding for the Green Fluorescent Protein) (Rai and Boyle 2007). The IP foundations of the Repository are thus rather shaky (Calvert 2012).

5.4 Anticipating and Guarding against a Public Backlash

But the most interesting issue evoked in "Beyond divisions" is that of a public backlash. The tensions that animate the field are seen as an obstacle to public confidence. If synthetic biologists cannot agree on common standards and regulations, how will the public ever agree with them? The public is seen as a force of dissolution mirroring the lack of internal cohesion of the synthetic biology community, an attitude that could be referred to as "synbiophobia-phobia": the phobia of public synbiophobia (Marris 2015).

Interestingly, many statements in the special issue suggest that the internal divisions of synthetic biology are not only seen as an obstacle, they are *seen as being seen* — or *gazed at*[26] — as an obstacle by a threatening 'Other' who may use it to *weaken* synthetic biology. First, the social scientist is identified as one such 'Other':

> The paucity of connection between these groups has provided ample fodder for sociologists of science (Nature editors 2014, p.133).[27]

Then, other potential 'opponents', such as the Convention for Biological Diversity are mentioned:

> Storm clouds are gathering on the horizon. Not everyone agrees that synthetic biology is a force for good, and that opposition has found its voice in a consultation for the global Convention for Biological Diversity. [That is why] it is now more vital than ever that synthetic biologists present a united front (Nature editors 2014, p.133).

Finally, the public is depicted in a similar manner as a major threat. Using a very agonistic, military vocabulary ("force for good," "united front"), the editors of *Nature* suggest that synthetic biology must stand united in order to *win* public

---

26 In Jacques Lacan's psychoanalysis (1978), gaze is the anxious moment when the subject experiences itself being seen by the Other.

27 Of course, these statements are diametrically opposite to those of social scientists who engage in studies *with* scientists rather than making studies *of* scientists (Calvert and Frow 2014), and tend to think of synthetic biologists as "epistemic partners" (Holmes and Marcus 2008) rather than as 'actors' (*versus* social scientists as 'spectators'). Contrary to traditional ELSI approaches, a new generation of social scientists refuses to consider that their contribution is restricted to 'downstream' implications, hence their plea to reconsider social and ethical reflection and intervention as being, as well, an *epistemic* task through which they contribute *with* synbio scientists to the orientation of research and innovation choices (Rabinow and Bennett 2007; Balmer et al. 2012; Calvert 2013; Bensaude Vincent 2013b; Balmer and Bulpin 2013; Calvert and Frow 2014).

acceptance. Thus, a kind of mirror relationship takes place between two dissuasive images that mutually reinforce each other: the disunity of the field and the public distrust of the field. In this mirror relationship, the public is referred to as 'one' public because it is unified as *the* major source of risk, the risk of a rejection of the technology, and such unified image of the public serves in turn to compel synthetic biologists to unity. The construal of synthetic biology as a unified field despite its allegedly pathological internal divisions and the construal of the public as a confident recipient despite its allegedly pathological recalcitrance are thus one and the same process.

Everything unfolds as if synthetic biology had sought to construct and unify its self-image by positing an 'outside image' beheld by 'the others': social scientists, the regulatory bodies, the NGOs, and the greatest 'Other': the 'general public' or global 'Other'. The public is seen as an external threat to the community and as a potential resource for building a united synthetic biology community, provided that its trust could be won. The public is omnipresent and even indispensable for welding the community, but provided it is represented as a "disembodied" public, as Claire Marris puts it (2015, p.90). This explains why, until now, it is mainly synthetic biologists themselves who have framed the relevant ethical questions into a checklist of preformatted issues: biosafety, biosecurity, 'playing God', intellectual property, and — of course — public engagement. So far, the fear of synbiophobia had been the main driver for engaging the public: hardly a good basis to spur an inclusive debate over the values and finalities of synthetic biology.

As Claire Marris (2015) compellingly argues, the promises of synthetic biology to deliver untold blessings to society, and its susceptibility to fearful publics are two narratives that have been simultaneously co-constructed and mutually reinforce each other. Instead of allowing people to form and confront their opinions over what kinds of research directions are the more valuable, the definition of what counts as valuable societal benefits and the ways in which synthetic biology might contribute to them have not been opened to public deliberation. The public, Marris analyzes, is framed as a dual entity — quite in the same manner that one speaks of 'dual-use technologies':

> The disembodied public that is conjured up during discussions among scientific and governmental elites is typically conceived of either as an uncommitted member of the public who has no particular interest in the field or as overcommitted activists who are portrayed as being intent on shutting down the field. When referring to the first category, the public (usually in the singular) is represented as a passive (unmobilised) and malleable entity, easily swayed by information. It is assumed that 'sensationalist' and 'unscientific' information from campaigning groups or the mass media will produce negative attitudes, and that adequate scientific information will

produce positive attitudes. Mobilised publics, referred to as 'activists', 'NGOs' or 'environmental groups' are, on the other hand, portrayed as having entrenched views, and as terrifically, and terrifyingly, active. They are portrayed as being able to easily influence media representations of synthetic biology, and thus also the attitudes of the unmobilised public (Marris 2015, p.90).

Despite the repeated calls for "science with and for society" (European Commission 2014), the old image of a communication gap between science and the public is still vivid (Bensaude Vincent 2001). The performative fiction here — resilient despite the watchwords of the European Commission — is not one of synthetic biology as a unified discipline, it is one of an increasing gap to be filled between science and the public that invites being addressed by presenting "a united front".

6 Conclusion

I have depicted the landscape of synthetic biology as an intrinsically and constitutively differentiated terrain. While the multifarious techno-epistemic cultures and meanings of synthetic biology could — and *should* — be appraised as a source of creative tensions for spurring healthy controversies, the diversity is despised as one reflecting mere pathological divisions because of the obsession of synthetic biologists to win public acceptance. Yet the effect is counterproductive: the attempts of synthetic biologists to overcome their divisions further contribute to an even greater division between synthetic biology and 'the' public. If the dream of unity could turn into a poisoned dream, what is there to do, then?

First, instead of unifying synthetic biology, its technologies and its 'correct' public image, it might be more desirable to highlight and publicly value the divergences that play out between the multiple techno-epistemic cultures it comprehends. Nurturing and valuing diversity is preferable in order to let the technology become open to citizen shaping and to cross-fecundation with the different values and aspirations of a diversity of publics. To this end, it is important to avoid positing a standard model of the public defined as a single entity by its disposition to confidence or to rejection. People are legitimately reluctant to see their role restricted to mere acceptation or rejection of technological solutions; they aspire to participate in the definition of problems (Thomas 2015).

Second, acknowledging the diversity of synbio's means and goals, and of the aspirations they convey, could be regarded as a constructive first step for enabling a more epistemologically and socially robust technoscience instead of encouraging the stereotype of synthetic biology as a "mad Frankenstein 2.0. science" mainly committed to biocapitalistic power and profit (Meyer et al. 2013). Doing so could support the view of synthetic biology as a pluralistic enterprise holding creative tensions instead of as a monolithic block converging on a promethean agenda based on a reductionist epistemology; and a view of researchers who, much like concerned citizens nowadays, are facing and debating different choices as to how and why our relationship to the nonhuman living world is transforming. The intrinsic diversity of synthetic biology is not something that should be overcome to allow for a more trustful relation with 'the' public (singular). It should be taken for granted and even enhanced for the sake of both a lively technoscience and a lively democratic debate among a plurality of 'publics' (plural). It is time to stop trying to bridge the gap and to try instead to *multiply* the gaps by recognizing that there is no such thing as a single and unique frontline between science and the public; the real, interesting, valuable, and mattering frontlines are those between different possible worlds. As Bruno Latour put it,[28] "the frontline is a 'worldline'" (*"la ligne de front est une ligne de monde"*). Many issues associated with synthetic biology, such as that of resources (and which resources for whom?) and issues concerning the designation and management of common goods and the legitimate means of trading with nonhuman living forms, outline different possible common worlds that the technoscience, the public and even nature would have to share and inhabit together.

Third, nurturing diversity in synthetic biology and its publics should allow possible choices of society to be identified, collectively discussed, and decided upon. Indeed, synthetic biology is the "biology of the why not";[29] it explores *possible* life forms which were not selected by natural evolution or which depend on other constraints and generating mechanisms. As such, the main epistemic character of synthetic biology is its opportunism: taking "whatever works" (Kastenhofer 2013) in an opportunistic exploration of the possible; multiplying possible life-forms and possible worlds. But are all possibilities *possible together*, able to exist at the same time and in the same common world, without choices to be made? Thus, the question of 'what works for synthetic biology and society in the long run' is not only a question of possibility; it is also, to use a Leibnizian concept, a question of *compossibility*—what is possible together?

> [N]ot all possibles are compossible. Thus, the universe is only a certain collection of compossibles, and the actual universe is the collection of all existing possibles, that is to say, those which form the richest composite. And since there are different com-

---

28 During Sara Angeli Aguiton's fascinating thesis defense (Angeli Aguiton 2014) at the Institut d'études politiques Sciences-Po Paris, December 15, 2014.
29 I borrow this phrase from Bernadette Bensaude Vincent.

binations of possibilities, some of them better than others, there are many possible universes, each collection of compossibles making up one of them (Leibniz 1960, p.662).

Be it on a biological, technological or political level, not everything is compossible or possible together for a diversity of human and nonhuman beings inhabiting a shared common world. Highlighting the incompossibilities of synthetic biology would allow the identification of possible choices of society, that is, choices between possible worlds. Thus, instead of unifying the field behind the flagship statement of 'making life easier to engineer' and applying the principle of 'whatever works', it might be more socially relevant to allow people to *contrast and compare* different approaches of synbio in order to learn to better situate the tipping points where choices of society—of possible and compossible worlds—are to be addressed.

Let us finish with a proposition concerning the elaboration of standards. Instead of packaging the whole field of synthetic biology into a big standardized black-box, why not engage people to participate in the elaboration of different standards suited to different technological as well as societal goals? Why not open the standards to social shaping, not only in technical terms, but also by working on the embodiment of a diversity of values in these standards?

Acknowledgments

This work contributes to the Mobilisation and Mutual Learning Action Plan (MMLAP) SYNENERGENE (*Synthetic biology – Engaging with New and Emerging Science and Technology in Responsible Governance of the Science and Society Relationship*), supported by the European Commission under the 7th Framework Programme "Science in Society". The SYNENERGENE workshop Worldviews & Values in Synthetic Biology held at the University Paris 1 Panthéon – Sorbonne on June 6-7, 2014 has provided meaningful resources for this article. I wish to express my gratitude to all the participants.

References

Angeli Aguiton, S. (2010). *Un vivant "sexy et à peu près faisable."* Anthropologie d'un concours d'ingénierie génétique. Master's Dissertation. École des Hautes Études en Sciences Sociales, Paris.

Angeli Aguiton, S. (2014). *La démocratie des chimères: gouvernement des risques et des critiques de la biologie synthétique, en France et aux Etats-Unis.* Thesis/Dissertation. Institut d'études politiques Sciences-Po Paris.

Annaluru, N., Muller, H., Mitchell, L.A., Ramalingam, S., Stracquadanio, G., Richardson, S.M., ... Chandrasegaran, S. (2014). Total synthesis of a functional designer eukaryotic chromosome. *Science, 344*(6179), 55-58. doi: 10.1126/science.1249252.

Attwater, J., & Holliger, P. (2014). A synthetic approach to abiogenesis. *Nature Methods, 11*, 495-498.

Balmer, A., Bulpin, K., Calvert, J., Kearnes, M., Mackenzie, A., Marris, C., Martin, P, Molyneux-Hodgson, S., & Schyfter, P. (2012). Towards a Manifesto for Experimental Collaborations between Social and Natural Scientists. http://experimentalcollaborations.wordpress.com/. Accessed: 12 December 2014.

Balmer, A.S., & Bulpin, K.J. (2013). Left to their own devices: Post-ELSI, ethical equipment and the International Genetically Engineered Machine (iGEM) Competition. *BioSocieties, 8*, 311-335. http://www.ncbi.nlm.nih.gov/pmc/articles/PMC3772706/. Accessed: 28 November 2014.

Batt, G. (2014). Cybergenetics: cells driven by computers. Communication at the workshop *Worldviews and Values in Synthetic Biology*, University Paris 1 Panthéon-Sorbonne, June 6-7, 2014. Written report at http://synenergene.eu/sites/default/files//uploads/WV%26Vs%20Summary%20Report%20of%20the%20Workshop_1.pdf (pp.5-6). Accessed: 28 November 2014.

Bedau, M.A. (2014). How weak emergence drives synthetic biology. Communication at the workshop *Worldviews and Values in Synthetic Biology*, University Paris 1 Panthéon-Sorbonne, June 6-7, 2014. Written report at http://synenergene.eu/sites/default/files//uploads/WV%26Vs%20Summary%20Report%20of%20the%20Workshop_1.pdf (pp.13-14). Accessed: 28 November 2014.

Bensaude Vincent, B. (2001). A genealogy of the increasing gap between science and the public. *Public Understanding of Science, 10*, 99-113.

Bensaude Vincent, B. (2013a). Discipline-building in synthetic biology. *Studies in History and Philosophy of Science Part C: Studies in History and Philosophy of Biological and Biomedical Sciences, 44*, 122-129.

Bensaude Vincent, B. (2013b). Ethical perspectives on synthetic biology. Biological Theory, Massachusetts Institute of Technology Press (MIT Press): STM Titles, 2013, 8(4), 368-375. <hal-00937211>.

Benner, S.A. (2004). Redesigning genetics. *Science, 306*, 625-626.

Benner, S.A., & Sismour, A.M. (2005). Synthetic biology. *Nature Reviews Genetics, 6*, 533–543.

Benner, S.A., Chang F., & Yang Z. (2011). Synthetic biology, tinkering biology, and artificial biology: A perspective from chemistry. In: P. L. Luisi, & C. Chiarabelli (2011), *Chemical Synthetic Biology* (pp.69-106). Chichester, UK: John Wiley & Sons.

Calvert, J. (2012). Ownership and sharing in synthetic biology: A "diverse ecology" of the open and the proprietary? *BioSocieties, 9*, 169-187.

Calvert, J. (2013). Collaboration as a research method? Navigating social scientific involvement in synthetic biology. In: N. Doorn, D. Schuurbiers, I. van de Poel, M. E. Gorman (eds.),

Early engagement and new technologies: opening up the laboratory (pp.175-194). Dordrecht, Heidelberg, New York, London: Springer.

Calvert, J., & Frow, E. (2014). Experimental Collaborations. Communication at the *Workshop on the Research Agendas in the Societal Aspects of Synthetic Biology*, Arizona State University, November 4-6, 2014.

Campos, L. (2009). That was the synthetic biology that was. In: M. Schmidt, A. Kelle, A. Ganguli-Mitra, & H. de Vriend (eds.), *Synthetic biology: The technoscience and its consequences* (pp.5-21). Dordrecht: Springer.

Carlson, R.H., (2011). *Biology Is Technology. The Promise, Peril, and New Business of Engineering Life*. Harvard: Harvard University Press.

Chan, L.Y., Kosuri, S., & Endy, D. (2005). Refactoring bacteriophage T7. *Molecular Systems Biology, 1*, 2005.0018. doi: 10.1038/msb4100025.

Church, G.M., Elowitz, M.B., Smolke, C.D., Voigt, C.A., & Weiss, R. (2014). Realizing the potential of synthetic biology. *Nature Reviews Molecular Cell Biology, 15*, 289-294.

Collins, J.J. (2014). Synthetic biology: How best to build a cell? Bring in the biologists. Comments. *Nature, 509*, 155–156.

Delaplace, F. (2014). On the use of computer language for programming biological function. Communication at the workshop *Worldviews and Values in Synthetic Biology*, University Paris 1 Panthéon-Sorbonne, June 6-7, 2014. Written report at http://synenergene.eu/sites/default/files//uploads/WV%26Vs%20Summary%20Report%20of%20the%20Workshop_1.pdf (pp.21-22). Accessed: 27 November 2014.

Delcourt, M. (2014). Global Bioenergies – An application of synthetic biology: from carbohydrates to hydrocarbons. Communication at the workshop *Worldviews and Values in Synthetic Biology*, University Paris 1 Panthéon-Sorbonne, June 6-7, 2014. Written report at http://synenergene.eu/sites/default/files//uploads/WV%26Vs%20Summary%20Report%20of%20the%20Workshop_1.pdf (pp.6-7). Accessed: 29 November 2014.

Gardner, T.S., Cantor, C.R., & Collins, J.J. (2000). Construction of a genetic toggle switch in Escherichia coli. *Nature, 403*, 339-342.

Ellington, A. (2014). Synthetic biology: How best to build a cell? Build green "bioalchemists." Comments. *Nature, 509*, 155–156.

Elowitz, M.B., & Leibler, S. (2000). A synthetic oscillatory network of transcriptional regulators. *Nature, 403*, 335-338.

Endy, D. (2005). Foundations for engineering biology. *Nature, 438*, 449-453.

Engelhard, M. (2014). Differentiating the evaluation of synthetic biology. Communication at the workshop *Worldviews and Values in Synthetic Biology*, University Paris 1 Panthéon-Sorbonne, June 6-7, 2014. Written report at http://synenergene.eu/sites/default/files//uploads/WV%26Vs%20Summary%20Report%20of%20the%20Workshop_1.pdf (p.3). Accessed: 29 November 2014.

European Commission 2014. Science with and for society. Horizon 2020 Work Programme 2014-2015, part 16. http://ec.europa.eu/research/participants/data/ref/h2020/wp/2014_2015/main/h2020-wp1415-swfs_en.pdf. Accessed: 28 January 2014.

Frow, E., & Calvert, J. (2013). "Can simple biological systems be built from standardized interchangeable parts?" Negotiating biology and engineering in a synthetic biology competition. *Engineering Studies, 5*, 42-58.

Fussenegger, M. (2014). Synthetic biology: How best to build a cell? Make tools for mammalian cells. Comments. *Nature, 509*, 157.

Gibbons, M., Limoges C., Nowotny H., Schwartzman S., Scott P., & Trow M. (1994). *The New Production of Knowledge. The Dynamics of Science and Research in Contemporary Societies*. London: Sage.

Gabrys, J. (2013). Plastic and the work of the biodegradable. In: J. Gabrys, G. Hawkins, & M. Michael (eds.). *Accumulation: The Material Politics of Plastic* (pp.208-227). London & New York: Routledge.

Geim, A. K., & Novoselov, K. S. (2007). The rise of graphene. *Nature Materials*, 6, 183-191.

Henkel, J., & Maurer, S.M. (2009). Parts, property and sharing. *Nature Biotechnology, 27*, 1095-1098.

Hobom, B. (1980). Surgery of genes – at the doorstep of synthetic biology. *Medizinische Klinik, 75*, 14-21.

Holmes, D., & Marcus, G. (2008). Collaboration today and the re-imagination of the classic scene of fieldwork encounter. *Collaborative Anthropologies, 1*, 81-101.

Jaramillo, A. (2014). General-Purpose Programmable Evolution Machine on a Chip. Communication at the workshop *Worldviews and Values in Synthetic Biology*, University Paris 1 Panthéon-Sorbonne, June 6-7, 2014. Written report at http://synenergene.eu/sites/default/files//uploads/WV%26Vs%20Summary%20Report%20of%20the%20Workshop_1.pdf (pp.12-13). Accessed: 24 November 2014.

Jullien, L. (2014). Biology-assisted Chemistry as an attractive paradigm for chemistry. Communication at the workshop *Worldviews and Values in Synthetic Biology*, University Paris 1 Panthéon-Sorbonne, June 6-7, 2014. Written report at http://synenergene.eu/sites/default/files//uploads/WV%26Vs%20Summary%20Report%20of%20the%20Workshop_1.pdf (pp.7-8). Accessed: 29 November 2014.

Kastenhofer, K. (2013). Two sides of the same coin? The (techno)epistemic cultures of systems and synthetic biology. *Studies in History and Philosophy of Biological and Biomedical Sciences, 44*, 130-140.

Lacan, J. (1978). *Seminar XI: The Four Fundamental Concepts of Psychoanalysis*. New York, London: Norton and Co.

Lacey, H. (2012). Reflections on science and technoscience. *Scientiae Studia, 10*, 103-128.

Lacour, S. (2011). A legal version of the nanoworld. *Comptes-rendus de l'Académie des Sciences de physique, 12*, 693-701.

Leibniz, G.W. (1960). *Die Philosophische Schriften von Gottfried Wilhelm Leibniz*, vol. 3 (First ed. C. I. Gerhardt. Berlin: Weidmann, 1875-90). Hildesheim: Olms.

Lentini, R., Santero, S.P., Chizzolini, F., Cecchi, D., Fontana, J., Marchioretto, M., ... Mansy, S.S. (2014). Integrating artificial with natural cells to translate chemical messages that direct E. coli behaviour. *Nature Communications, 5*, 4012. doi: 10.1038/ncomms5012.

Levskaya, A., Chevalier, A.A., Tabor, J.J., Simpson, Z.B., Lavery, L.A., Levy, M., ... Voigt, C.A. (2005). Synthetic biology: Engineering Escherichia coli to see light. *Nature, 438*(7067), 441-442.

Loeve, S. (2009). *Le concept de technologie à l'échelle des molecules-machines. Philosophie des techniques à l'usage des citoyens du nanomonde*. Thesis/Dissertation. University Paris Ouest Nanterre – La défense.

Loeve, S. (2014). Summary Report of the Paris Workshop *Worldviews and Values in Synthetic Biology*. Synenergene. http://synenergene.eu/resource/summary-report-paris-workshop-worldviews-and-values-synthetic-biology. Accessed: 5 May 2015.

Loeve, S. (2014). Introductory remarks to the session "Ways of doing synthetic biology." Communication at the Workshop *Worldviews and Values in Synthetic Biology*, University

Paris 1 Panthéon-Sorbonne, June 6-7, 2014. Written report at http://synenergene.eu/sites/default/files//uploads/WV%26Vs%20Summary%20Report%20of%20the%20Workshop_1.pdf (pp.4-5). Accessed: 29 November 2014.

Luisi, P.L. (1997). About various definitions of life. *Origins of Life and Evolution of the Biosphere, 28*, 613-622.

Luisi, P.L. (1998). Does science see a purpose in life? Communication at the workshop *Is There a Purpose in Nature? How to Navigate Between the Scylla of Mechanism and Charybdis of Teleology*. Charles University Prague, March 22-25, 1998. http://web.archive.org/web/20070611152326/http://www.cts.cuni.cz/conf98/luisi.htm. Accessed: 19 January 2015.

Luisi, P.L. (2006). *The Emergence of Life: From Chemical Origins to Synthetic Biology*. Cambridge: Cambridge University Press.

Luisi, P.L. (2011). The synthetic approach in biology: epistemological notes for synthetic biology. In: P. L. Luisi & C. Chiarabelli (2011). *Chemical Synthetic Biology* (p. 343-362). Chichester, UK: John Wiley & Sons.

Mackenzie, A. (2013a). Realizing the promise of biotechnology: Infrastructural-icons in synthetic biology. *Futures, 48*, 5-12.

Mackenzie, A. (2013b). Synthetic biology and the technicity of biofuels. *Studies in History and Philosophy of Biological and Biomedical Sciences, 44*, 190-198.

Malaterre, C. (2013). Synthetic biology and synthetic knowledge. *Biological Theory, 8*, 346-356.

Mansy, S.S. (2014). Building artificial cells to deceive natural cells. Communication at the Workshop *Worldviews and Values in Synthetic Biology*, University Paris 1 Panthéon-Sorbonne, June 6-7, 2014. Written report at http://synenergene.eu/sites/default/files//uploads/WV%26Vs%20 Summary%20Report%20of%20the%20Workshop_1.pdf (pp.23-24). Accessed: 29 November 2014.

Marcovich, A., & Shinn, T. (2012). Regimes of science production and diffusion: towards a transverse organization of knowledge. *Scientiae Studia, 10*, 33-64.

Marcovich, A., & Shinn, T. (2014). *Towards a New Dimension: Exploring the Nanoscale*. Oxford: Oxford University Press.

Marliere, P. (2009). The farther, the safer: a manifesto for securely navigating synthetic species away from the old living world. *Systems and Synthetic Biology, 3*, 77-84.

Marris, C. (2014). The Construction of Imaginaries of the Public as a Threat to Synthetic Biology. *Science as Culture, 24*, 83-98.

Maxon, M. (2014). Synthetic biology: How best to build a cell? Agree on a definition. Comments. *Nature, 509*, 156.

Meyer, A., Cserer, A., & Schmidt, M. (2013). Frankenstein 2.0.: Identifying and characterising synthetic biology engineers in science fiction films. *Life Sciences Society and Policy, 9*, 1-17.

Morange, M. (2009). A new revolution? The place of systems biology and synthetic biology in the history of biology. *EMBO Reports, 10*, S50-S53.

Nature (2014). Synthetic Biology: Beyond Divisions. *Nature, 509*.

Nature editors (2014). Tribal gathering. *Nature, 509*, 133.

Nature Methods editors (2014). Synthetic biology: back to the basics. *Nature Methods, 11*, 463.

Nature Nanotechnology editors (2006a). Small is different. *Nature Nanotechnology, 1*, 1.

Nature Nanotechnology editors (2006b). nan'o·tech·nol'o·gy n. *Nature Nanotechnology, 1*, 8-10.

Nature Publishing Group (2014). Nature Special – Synthetic Biology: Beyond Divisions. <Nature.Publishing.Group@ealert.nature.com>, 8 May 2014 4:06 pm.

Nature Reviews Microbiology editors (2014). Milestones in synthetic (micro)biology. *Nature Reviews Microbiology, 12*, 309.

Novoselov, K.S., Fal, V.I., Colombo, L., Gellert, P.R., Schwab, M. G., & Kim, K. (2012). A roadmap for graphene. *Nature, 490*(7419), 192-200. doi: 10.1038/nature11458.

Novoselov, K.S., Geim, A. K., Morozov, S. V., Jiang, D., Katsnelson, M. I., Grigorieva, I. V., ... Firsov, A. A. (2005). Two-dimensional gas of massless Dirac fermions in graphene. *Nature, 438*(7065), 197-200. doi: 10.1038/nature04233.

Novoselov, K.S., Geim, A. K., Morozov, S. V., Jiang, D., Zhang, Y., Dubonos, S. V., ... Firsov, A. A. (2004). Electric field effect in atomically thin carbon films. *Science, 306*(5696), 666-669. doi: 10.1126/science.1102896.

O'Malley, M.A. (2009). Making knowledge in synthetic biology: Design meets kludge. *Biological Theory, 4*, 378-389.

O'Malley, M.A., Powell, A., Davies, J. F., & Calvert, J. (2008). Knowledge-making distinctions in synthetic biology. *BioEssays, 30*, 57-65.

Pastrana, E. (2011). Optogenetics: controlling cell function with light. *Nature Methods, 8*, 24-25.

Peplow, M. (2013). Graphene: The quest for supercarbon. *Nature, 503*, 327-329.

Pompon, D. (2014). Frontiers and views on metabolic engineering. Communication at the Workshop *Worldviews and Values in Synthetic Biology*, University Paris 1 Panthéon-Sorbonne, June 6-7, 2014. Written report at http://synenergene.eu/sites/default/files//uploads/WV%26Vs%20Summary%20Report%20of%20the%20Workshop_1.pdf (pp.20-21). Accessed: 29 November 2014.

Potthast, T. (2009). Paradigm shifts versus fashion shifts? *EMBO Reports, 10*, S42-S45.

Rabinow, P., & Bennett, G. (2007). From bioethics to human practices, or assembling contemporary equipment. In B. da Costa & K. Philip (eds.), *Tactical biopolitics art, activism, and technoscience* (pp.389-400). Cambridge: MIT Press.

Rai, A., & Boyle, J. (2007). Synthetic biology: caught between property rights, the public domain, and the commons. *PLOS Biology, 5*, e58.

Raimbault, B., Cointet, J.P., & Joly, P.B. (2013). Caractérisation du processus d'émergence de la biologie synthétique à partir d'une approche scientométrique. *Médecine/Sciences, 29*, 47-55.

Rawls, R.L. (2000). "Synthetic biology" makes its debut. *Chemical Engineering News, 78*, 49-53.

Roco, M.C., & Bainbridge, W. S. (2002). Converging technologies for improving human performance: integrating from the nanoscale. *Journal of Nanoparticle Research, 4*, 281-295.

Römer, W. (2014). Synthetic membrane biology – rebuilding cellular processes on artificial membrane systems. Communication at the Workshop *Worldviews and Values in Synthetic Biology*, University Paris 1 Panthéon-Sorbonne, June 6-7, 2014. Written report at http://synenergene.eu/sites/default/files//uploads/WV%26Vs%20Summary%20Report%20of%20the%20Workshop_1.pdf (p.25). Accessed: 29 November 2014.

Sauro, H. (2014). Synthetic biology: How best to build a cell? Capitalize on evolution. Comments. *Nature, 509*, 157.

Schmidt, M. (ed.) (2012). *Synthetic biology: industrial and environmental applications*. Weinheim: John Wiley & Sons.

Szostak, J.W., Bartel, D.P., & Luisi, P.L. (2001). Synthesizing life. *Nature, 409*, 387-390.

Thomas, J. (2015). Constructing a "futurology from below": a civil society contribution toward a research agenda. *Journal of Responsible Innovation*, (ahead-of-print), 1-4.

UK Synthetic Biology Roadmap Coordination Group (2012). *A Synthetic Biology Roadmap for the UK* http://www.rcuk.ac.uk/RCUK-prod/assets/documents/publications/SyntheticBiologyRoadmap.pdf. Accessed: 20 December 2014.

Weiss, R. (2014). Synthetic biology: How best to build a cell? Automate efficient design. Comments. *Nature, 509*, 157.

Synthetic Genome Technologies

Christoph Then

Summary

The current debate on synthetic biology and artificial life is often seen as a theoretical exercise with not much basis in reality. So far, laboratories have not made many advances in creating completely new life forms, but at the same time, the increasing technical potential to radically change the genome of existing life forms has escaped broader public attention.

Presently, synthetic biology appears to be emerging as a much more realistic and powerful tool to manipulate genomes in existing organisms and is outpacing concepts to create 'new life forms'. These developments will trigger a whole new set of specific issues and questions on the ethical consequences, social impacts, and environmental risks.

The discussion in this paper is focused on technologies relating to DNA, bearing in mind that synthetic biology also includes experiments on cell systems. Thus the term *synthetic genome technologies* is used in this paper to underscore that these new technologies are strongly linked to the methods used in genetic engineering and to those used in synthetic biology. *Genome editing* is another widely used term with a similar meaning.

The most relevant topics in this context are DNA synthesis, so-called gene scissors, and usage of oligonucleotides. The paper also includes some examples of applications in food plants, mammals, and insects as well as a discussion on potential usage in humans.

Several of the problems that we have identified concern issues not sufficiently covered by existing regulations. For example:

- It is very important to protect existing life forms, but equally important to protect the future evolution of life / biodiversity;

- The genetic integrity of existing life is a relevant issue even if animals do not suffer or the environment is not at risk;
- New approaches to regulate access to specific technologies or genetic information have to be considered if such technologies can be used to produce hazardous organisms.

1 Introduction

There have been several attempts within the field of synthetic biology to create artificial 'new life forms', but it appears that none of these attempts has been successful so far. For example, in 2007 Craig Venter, one of the most well-known experts in this field, announced that an application for a new organism, based on a minimal bacterial genome, had been submitted to the patent office (International Patent WO 2007/047148, 2007). However, as far as we know, this life form has not been established in the laboratories, and in 2010 Venter and his team presented another micro-organism which was able to reproduce and possessed a genome that was wholly synthesized but in essence not new (Gibson et al. 2010).

A different perspective on synthetic biology from the one emerging from Venter's research has been described by George Church and Ed Regis in their book *Regenesis – How Synthetic Biology Will Reinvent Nature and Ourselves* (Church and Regis 2012). Church is a prominent researcher in the field of synthetic biology. He puts more emphasis on technologies that enable radical changes in the genome of *existing* life forms than on the creation of completely *new* organisms. According to Church and Regis, new technologies will enable the rewriting of large parts of the genome – including the human genome:

> The same technique would work for the Neanderthal, except that you'd start with a stem cell genome from a human adult and gradually reverse-engineer it into the Neanderthal genome or a reasonable close equivalent. [...] If society becomes comfortable with cloning and sees value in true human diversity, the whole Neanderthal creature itself could be cloned by a surrogate mother chimp – or by an extremely adventurous female human (Church and Regis 2012).

This perspective naturally triggers a different set of questions about ethical consequences, social impacts, and environmental risks than the debate focusing on the invention of 'new life forms'. Taking this perspective as a starting point, the present paper gives a short overview of some crucial methods and relevant applications. We have, in this context, introduced a slightly different terminology to the one commonly used: in light of the fact that the technologies discussed in our paper have a strong focus on DNA, and bearing in mind that synthetic biology also includes experiments on cell systems, the term *synthetic genome technologies* is used in this paper to make it clear that these new technologies are strongly linked to the methods used in genetic engineering as well as to those used in synthetic biology (see also Then 2015). *Genome editing* is another widely used term with a similar meaning.

2   Synthetic Genome Technologies – methods and applications

In 2010, a microorganism with a completely resynthesized genome was presented to the public for the first time (Gibson et al. 2010). The media release from the Craig Venter Institute reads:

> [This] is the proof of principle that genomes can be designed in the computer, chemically made in the laboratory and transplanted into a recipient cell to produce a new self-replicating cell controlled only by the synthetic genome (JCVI 2010).

Thus this experiment should largely be seen as proof of concept that organisms can propagate on the basis of a synthetic genome. As mentioned earlier, synthetic biology is able not only to resynthesize the DNA of existing life forms, but also to radically alter their genomes. In March 2014, it became public that scientists in the US had succeeded in resynthesizing a whole chromosome contained in yeast. In the process, they removed large parts of the genome regarded as non-essential for survival of the yeast cell (Annaluru et al. 2014). Apparently, the newly available methods are very different to the methods currently known among the wider public as genetic engineering. Some characteristics of these new technologies are:

- It is not necessary to isolate DNA from living beings; the DNA can be synthesized directly in the laboratory.
- The structure of the DNA does not depend on naturally existing genomes, it can be designed in the laboratory without a native template or be a combination of DNA from various organisms.
- Some applications enable direct alteration of the genome in the cells, without isolating or transferring DNA.

- At the same time, the technical possibilities for changing the regulation of the natural genome without changing the structure of the DNA are becoming increasingly important.

Methods of analyzing and synthesizing DNA are described elsewhere (e.g., Gibson et al. 2010). In this paper we describe two methods that are used in synthetic genome technologies to achieve far-reaching changes in the genome and have so far attracted little public attention. These methods consist of so-called genome scissors (nucleases) and oligonucleotides (shortcuts of DNA or RNA).

2.1 Some relevant methods

Genome scissors (nucleases)

Nucleases are proteins (enzymes) used to break up DNA – that is why they are called genome scissors. Nucleases have been technically available for several years but were previously limited in terms of number of genome-cutting options. Several nucleases which allow the cutting of DNA and the insertion of new DNA in any position in the chromosomes have been developed over the last few years. Some of these new genome scissors are known as TALENs (Transcription Activator-Like Effector Nucleases) or CRISPR (Clustered Regularly Interspaced Short Palindromic Repeats). They consist of a combination of a unit to recognize specific regions of the DNA and an enzyme to cut the DNA. By using TALENs or CRISPR, genes can be knocked out (silenced), and mutations or new DNA can be inserted. A number of publications show that the new technologies might trigger new risks. For example, an investigation using human cells (Fu et al. 2013) shows that CRISPR can cause unintended mutations in the genome.

Oligonucleotides

Oligonucleotide Directed Mutagenesis (ODM) technology is based on using small parts of DNA (RNA) sequences (called oligonucleotides) which are synthesized according to naturally occurring templates (e.g., from plants). Synthesizing the plant's DNA alters one detail of it; for example, it can be changed to make the plant resistant to specific herbicides. These short synthetic sequences are transferred into the plant cell to induce alteration of the DNA at the region where the original template was sequenced. If successful, the native DNA of the plant will be altered accordingly. It is assumed that the oligonucleotides do not become integrated into the plant's DNA, but the detailed mechanisms are not known. It is believed that natural repair mechanisms, for example, can be the reason for the plant to adapt its genome to the synthetic DNA (Lusser et al. 2011).

At first sight it seems oligonucleotide technology as described can be used to achieve similar results as, for example, in mutation breeding of plants. However, on closer examination it is evident that it is to some extent also fundamentally different to mutation breeding. In conventional mutation breeding, an unspecific stress triggers non-targeted changes in the plant's DNA. Manipulation with oligonucleotides, however, is an intentional and invasive method that aims to change the DNA in a very specific way. Its overall impact on the genome is therefore also likely to be different. It is known that oligonucleotide technology can cause off-target effects (see, e.g., Vogel 2012; Pauwels et al. 2013). So far, there has been no systematic investigation into whether these effects and their patterns are different in comparison with those observed in mutation breeding.

Oligonucleotide technology can also be used to change longer sequences of DNA if applied repeatedly, as is the case, for example, in so-called *Multiplex Automated Genome Engineering* or *MAGE* (Carr et al. 2012). This method can be compared to an assembly line that goes round in circles with many workers introducing small changes every time the cell passes around. The higher the number of times it passes around, the higher the degree of alteration. According to Church, this technology can be used to transform the genome of an elephant into the genome of a mammoth, or the genome of homo sapiens into the genome of a Neanderthal (Church and Regis 2010).

2.2   Some current applications

Several publications show that technologies like TALENs and CRISPR are being applied to all kinds of organisms. Preliminary studies detailing the application of these techniques to plants (such as Arabidopsis, sorghum, rice, and wheat), fish, flies, worms, rats, rabbits, frogs, non-human primates and human cells (Sander and Joung 2014) and also to animals used in food production such as cattle (Tan et al. 2013), sheep (Han et al. 2014) and pigs (Hai et al. 2014) suggest that technologies such as CRISPR and TALENs are universally applicable across the biological kingdoms, including humans. They enable targeted DNA manipulation, even at multiple sites simultaneously (Bortesi and Fischer 2014; Segal and Meckel 2013; Baker 2014). Based on the CRISPR system, researchers developed what they called the *mutagenic chain reaction* (MCR). As shown for fruit flies, this technique enables newly inserted mutations to efficiently spread from one chromosome of origin to the homologous chromosome, thereby converting heterozygous mutations to homozygosity in the vast majority of somatic and germline cells. Once released, this new genetic information could easily be spread with native populations (Gantz and Bier 2015).

Several other applications derived from synthetic genome technologies have already been commercialized or might be allowed onto the market soon. Examples include the following:

- SmartStax maize is a joint Monsanto and Dow AgroSciences product. The maize produces six insecticidal toxins to kill pest insects that feed on the plants and is resistant to two herbicides. At least one of the toxins it produces, Cry1A.105, is based on a fusion of DNA which does not have a natural template.
- The genome of genetically engineered olive flies produced by the British company Oxitec is a synthetic combination of DNA from other insects, marine organisms, bacteria, and viruses. It is intended to be introduced in countries like Spain to combat economic damage caused in olive production by the larvae of the flies that live in the olives. When the genetically engineered flies mate with native flies a lethal gene will be introduced into natural populations, killing the females while the male offspring survive and spread their deadly genetic trait.
- The US company Intrexon relies on the methods used in synthetic genome technologies to radically change the genome of mammals and other organisms. Intrexon sees itself as "a leader in synthetic biology." According to the Intrexon website, their business is about to provide comprehensive genetic control of the biological functions of all kinds of species:

    Intrexon Corporation is [...] focused on the industrial engineering of synthetic biology [...] across multiple industry sectors, including: human therapeutics, protein production, industrial products, agricultural biotechnology, and animal science. The company's advanced bioindustrial engineering platform enables [...] unprecedented control over the function and output of living cells (Intrexon 2013).

    Intrexon files patents on all kinds of mammals, including chimpanzees manipulated with synthetic DNA stemming from insects.

- There are several companies, including Charles River, that produce animal models for pharmaceutical companies. They supply mice and rats manipulated with technologies such as TALENs called "fast knock-in animal models." The animals are produced on demand as "custom engineered rodents." Mice or rats can be created within only five months by the insertion of any DNA at a targeted location ("knock-in"). These are fast-track, man-made creations of animals inheriting digitally designed DNA.

- We might have humans with synthetic DNA much sooner than the resurrection of the Neanderthal: for example, Inovio Pharmaceuticals are developing DNA vaccines to be inserted into human cells to trigger specific immune responses.

3 Questions for further discussion

Clearly, there are several issues that are relevant in terms of regulatory oversight, including risks for the environment and ethical questions.

For example, we cannot predict the long-term ecological consequences of releasing organisms with radically altered genomes that escape spatio-temporal control. There is a great deal of uncertainty surrounding future environmental conditions and the long-term characteristics of such organisms. In evolutionary processes it is quite possible that events with a low probability of occurring may begin to acquire a reasonable chance of occurring. Lack of knowledge in too many areas means the necessary prerequisites for reliable risk assessment do not exist (Bauer-Panskus et al. 2013). Consequently, synthetic organisms whose spatio-temporal dimension cannot be controlled should not be released, no matter for what purpose they were made.

Applications such as those for the olive fly raise general questions regarding our responsibility for biodiversity: Should we introduce lethal genes into native populations if they might not only suppress the number of individuals but also – in theory – change the germline of a whole species?

On the level of biosecurity, there is a growing need to discuss regulatory oversight of DNA synthesis and access to genetic information. Since DNA synthesis is becoming increasingly cheaper and faster, the risk of proliferation of new hazardous organisms is a substantial issue.

Further, there is an urgent need for debate on ethical boundaries. For example, synthetic genome technologies that enable substantial changes in the genetic identity of mammals can raise ethical concerns even if suffering and pain cannot be observed. Issues such as genetic identity and genetic integrity, which have hardly been discussed at all to date, are becoming increasingly relevant. Addressing these topics is also becoming an increasingly burning issue for the human species: In the March 2015 edition of the international magazine Science and Nature, researchers already called for a moratorium on genome editing in human embryos. They warn that genome-editing technologies such as CRISPR-Cas9 and TALENs make it easy for anyone with basic molecular biology training to insert, remove, and manipulate genes in cells—including sperm, eggs, and embryos, not only potentially curing genetic diseases but also adding new genetic characteristics for any other purposes (Vogel 2015).

Considering our own origins and the origin of all existing life may prove helpful in gaining a broader perspective and a better understanding of the crossroads we are at: all existing life forms emerged from first early cells. Life in its existing forms, but also in terms of its future evolution, is still in a continuum with its origin some billion years ago. As the philosopher Karl Popper put it:

> The first cell is still living after billions of years, and now even in many trillions of copies. Wherever we look, it is there. It has made a garden of our earth and transformed out atmosphere with green plants. And it created our eyes and opened them to the blue sky and the stars. It is doing well (Popper 1992, p.17).

We now have far-reaching technical possibilities to interfere with the "first cell" in its existing and future forms on the level of the genome. Such possibilities cover all kinds of living beings, including humans. We can create cells that are substantially different from the "first cell". We can create life that interferes with, disturbs, or interrupts the network and dynamics of current biodiversity and its further development. The future of life can be impacted by economic interests and technological failures to an extent not previously known. In his book, Popper explains that all life is "a search for a better world". Looking at the current options emerging from synthetic genome technologies, it seems a good idea to give the "first cell" a realistic chance to keep on searching for "a better world" by following its own patterns, dynamics, and networks, which we still do not understand. We should not try to design future life in the laboratory. We should treat life with respect and not with technical arrogance.

4   Conclusions

Current debates on synthetic biology and artificial life are often seen as a theoretical exercise. Actual developments on the level of synthetic genome technologies have so far largely escaped broader public attention. Several of the problems that have been identified concern issues that are not sufficiently covered by existing regulations. For example:

- It is becoming more and more important to protect not only existing life forms but also the future evolution of life / biodiversity;

- The genetic identity and integrity of existing life is a relevant issue even if environmental risks are not apparent and animals do not suffer as a result;
- New approaches to regulate access to specific technologies or genetic information need to be considered.

References

Annaluru, N., Muller, H., Mitchell, L.A., Ramalingam, S., Stracquadanio, G., Richardson, S.M., . . . Chandrasegaran, S. (2014). Total Synthesis of a Functional Designer Eukaryotic Chromosome. *Science, 344*(6179), 55-58. doi: 10.1126/science.1249252.

Baker, M. (2014). Gene editing at CRISPR speed. *Nature Biotechnology, 32*(4), 309-312.

Bauer-Panskus, A., Breckling, B., Hamberger, S., & Then, C. (2013). Cultivation-independent establishment of genetically engineered plants in natural populations: current evidence and implications for EU regulation. *Environmental Sciences Europe, 25*(1), 34. doi: 10.1186/2190-4715-25-34.

Bortesi, L., & Fischer, R. (2014). The CRISPR/Cas9 system for plant genome editing and beyond. *Biotechnology Advances, 33*(1), 41-52. doi: 10.1016/j.biotechadv.2014.12.006.

Carr, P.A., Wang, H.H., Sterling, B., Isaacs, F.J., Lajoie, M.J., Xu, G., Church, G.M., & Jacobson, J.M. (2012). Enhanced multiplex genome engineering through co-operative oligonucleotide co-selection. *Nucleic Acids Research. 40*(17), e132. doi: 10.1093/nar/gks455.

Church, G., & Regis, E. (2012). *Regenesis: How Synthetic Biology Will Reinvent Nature and Ourselves.* New York: Basic Books.

Fu, Y., Foden, J.A., Khayter, C., Maeder, M.L., Reyon, D., Joung, J.K., & Sander, J.D. (2013). High-frequency off-target mutagenesis induced by CRISPR-Cas nucleases in human cells. *Nature Biotechnology, 31*(9), 822-826.

Gantz, V.M., & Bier, E. (2015). The mutagenic chain reaction: A method for converting heterozygous to homozygous mutations. *Science, 348*(6233), 442-444.

Gibson, D.G., Glass, J.I., Lartigue, C., Noskov, V.N., Chuang, R.Y., Algire, M.A., . . . Venter, J.C. (2010). Creation of a bacterial cell controlled by a chemically synthesized genome. *Science, 329*(5987), 52-56.

Glass, J.T., Smith, H.O., Hutchinson, C.A., Alperovich, N.Y., & Assad-Garcia, N. (2007). *International Patent WO2007/047148.* Washington, D.C.: World Intellectual Property Organization.

Hai, T., Teng, F., Guo, R., Li, W., & Zhou, Q. (2014). One-step generation of knockout pigs by zygote injection of CRISPR/Cas system. *Cell research, 24*(3), 372-375.

Intrexon (2013). *Intrexon Corporation Raises $150 Million for Synthetic Biology Initiatives.* Intrexon Corporation. http://investors.dna.com/phoenix.zhtml?c=249599&p=irol-newsArticle&ID=1844278&highlight=. Accessed: 30 April 2015.

JCVI (2010). J. Craig Venter Institute. *First Self-Replicating Synthetic Bacterial Cell* [Press Release]. http://www.jcvi.org/cms/press/press-releases/full-text/article/first-self-replicating-synthetic-bacterial-cell-constructed-by-j-craig-venter-institute-researcher/. Accessed: 28 April 2015.

Lusser, M., Parisi, C., Plan, D., & Rodríguez-Cerezo, E. (2011). *New plant breeding techniques. Stateof-the-art and prospects for commercial development.(= JRC Scientific and Technical Reports/EUR 24760 EN)*.

Pauwels, K., Podevin, N., Breyer, D., Carroll, D., & Herman, P. (2014). Engineering nucleases for gene targeting: safety and regulatory considerations. *New Biotechnology, 31*(1), 18-27.

Popper, K. (1992). *In search of a better world*. London: Routledge.

Sander, J.D., & Joung, J.K. (2014). CRISPR-Cas systems for editing, regulating and targeting genomes. *Nature Biotechnology, 32*(4), 347-355.

Segal, D.J., & Meckler, J.F. (2013). Genome engineering at the dawn of the golden age. *Annual Review of Genomics and Human Genetics, 14*, 135-158.

Tan, W., Carlson, D.F., Lancto, C.A., Garbe, J.R., Webster, D.A., Hackett, P.B., & Fahrenkrug, S.C. (2013). Efficient nonmeiotic allele introgression in livestock using custom endonucleases. *Proceedings of the National Academy of Sciences, 110*(41), 16526-16531.

Then, C. (2015). *Handbuch Agrogentechnik*. München: Oekom Verlag.

Vogel, B. (2012). *Neue Pflanzenzuchtverfahren – Grundlagen für die Klärung offener Fragen bei der rechtlichen Regulierung neuer Pflanzenzuchtverfahren*. Bundesamt für Umwelt (BAFU), Sektion Biotechnologie, Bern; Baudirektion des Kantons Zürich, Amt für Abfall, Wasser, Energie und Luft (AWEL), Sektion Biosicherheit (SBS). www.awel.zh.ch/internet/baudirektion/awel/de/biosicherheit_neobiota/veroeffentlichungen/_jcr_content/contentPar/publication_2/publicationitems/titel_wird_aus_dam_e_0/download.spooler.download.1372927394124.pdf/Schlussbericht_NeuePflanzenzuchtverfahren_DEZ2012.pdf. Accessed: 30 April 2015.

Vogel, G. (2015). Embryo engineering alarm, researchers call for restraint in genome editing. *Science, 347*(6228), 1301. doi: 10.1126/science.347.6228.1301.

Promising applications of synthetic biology – and how to avoid their potential pitfalls

Bernd Giese, Henning Wigger, Christian Pade and Arnim von Gleich

1 Introduction

Synthetic biology is associated with great expectations as well as grave concerns. Currently, due to the small number of concrete applications, an assessment of innovations in synthetic biology has to start with an analysis of early indicators of both the opportunities and the risk-relevant features of expected applications. The field still remains largely in the basic research phase. Thus, we are rather forced to carry out *science* assessment instead of *technology* assessment (Paschen and Petermann 1992). But for that purpose the tools of assessment must be complemented with prospective approaches that have to be adapted to *technosciences* like synthetic biology (Nordmann 2011; Kastenhofer and Schmidt 2011).

Nevertheless, an initial analysis of certain basic features (i.e., functionalities) of structures, processes, and systems that are intended for future applications – not least in light of their hazard or exposure potential – could already prove to be sufficient for a classification of promising applications of synthetic biology.

2 Opportunities

The construction, or more precisely: the synthesis of biological structures – regardless of whether they are created as modified versions of already existing structures or as completely new creations – holds the opportunity of complementing the empirical methods of biological research with the practical experience of construction. And even if one is not convinced by Richard P. Feynman's famous statement that "What I cannot create, I do not understand," one must acknowledge that synthesis can be of great help to our understanding of biological structures and processes. In this sense, synthetic biology corresponds to synthetic chemistry and its contribution to the investigation of structure and function (Yeh and Lim 2007).

The characteristic methodology of rational construction and its approach, from the creation of a model to the layout of the design, right through to the construction of a final prototype, is of major importance with regard to fields of application. Such rational approach of synthetic biology, borrowed from classical engineering disciplines, promises to overcome the disadvantages of common biotechnology. Structures and systems commonly used in biotechnology have never completely stripped off a certain 'black box' character in the form of hidden interactions, owing to the complex cellular background with its plethora of (unwanted) functions, noise (in terms of variability of processes etc.), and an evolutionary influence. According to the common thesis, a rational approach could, instead, lead to transparency and predictability as well as shorter developmental periods due to the use of standardized parts or even orthogonal elements (Mutalik et al. 2013; Stanton et al. 2013).

The extent to which biological phenomena like noise and evolutionary changes can be integrated into the rational methodology of synthetic biology is of particular importance (Eldar and Elowitz 2010; Wang and Church 2011; Dymond et al. 2011). Also important is how genetic information can maintain specific functions stable over a number of generations (Canton et al. 2008). Perhaps classical engineering with its now rather complex constructions can even benefit from the experiences gained with synthetic biology in the context of complex systems.

Concentrating on the intended function and reducing all unnecessary complexity may be viewed as one essential advantage of synthetic biology. In this sense, a minimal cell should serve as a chassis for the implementation of genetic modules and devices to ensure their improved efficiency. However, a strategy that only focuses on genome reduction is viewed as controversial. According to experts in the field, emphasis should be placed on proper adaption of the host cell to the intended function, thus on specialization rather than on minimization alone (Dietz and Panke 2010). If the reduction of interfering cellular interactions is not only intended for efficient product synthesis or the quality of a required function, but in fact also used to remove risk-related functionalities, then applications could be designed to be safer. Strategies may include the appropriate reduction of already existing organisms (DeWall and Cheng 2011) or, as a quite extensive but safe approach, an in-vitro combination of biomolecules outside natural cellular contexts in solution, at surfaces or enclosed in vesicular compartments.[1]

---

1 An important step in this direction has been taken by the "Fraunhofer-Leitprojekt" on cell-free bioproduction, funded by the German Federal Ministry of Education and Research (BMBF) as part of the "Biotechnologie 2020+" program.

A promising application of synthetic biology is related to its analogy to synthetic chemistry, as mentioned above: progress in the construction of new metabolic pathways reveals that synthesis could shift from chemical processes to biotechnology (Khalil and Collins 2010). Nevertheless, a number of serious problems have to be solved because the exact interplay of the molecular components involved still requires time-consuming optimization (Kittleson et al. 2012). As the current example of the divestment of the biofuel company LS9 (cf. LaMonica 2014) has shown, ultimate decisions on investments in facilities for bioproduction depend in the end on the competitiveness of biosynthesis pathways with regard to established technological pendants, or to put it another way: commodity prices of state-of-the-art technology (fossil vs. biomass) (Khalil and Collins 2010).

Another option for synthetic biology would be to focus specifically on the problem of resources and raw materials. If synthetic biology were able to overcome its competition with food production by harnessing (industrial) waste materials such as syngas (Evonik 2013) instead of biomass (which so far is required for most biosynthesis technologies), a major problem of the so-called bio-economy would be solved.

Greater flexibility and extended variety of synthesis and metabolic conversions represent a valuable enrichment of the typical biotechnological methodologies and solutions. But with focus on metabolism, the real capabilities of living entities, apart from merely enzymatic functionalities, are left untapped. For example, in its potential to develop complex, hierarchically structured materials and tissues with an as yet unattained combination of technical qualities (cf. Vincent 2008; Cachat and Davies 2011; Laaksonen et al. 2011) synthetic biology offers a chance to overcome barriers in the field of biomimetics.

3   Potential pitfalls

To evaluate the applicability of beneficial approaches, their advantages have to be balanced against potential risks. A concrete investigation of benefits and especially risks depends on detailed knowledge about applications and application contexts, which is not available in the early stages of innovation (i.e., the research and development stage). In a prospective analysis we therefore have to look for indicators and clues. But what qualities of synthetic biological constructs might provide an indication of severe complications in application contexts that are not yet clearly defined? Here we propose that, as an initial orientation, a preliminary estimation of risk can rest on the two main factors for the emergence of risk: 1) hazardous functionalities and 2) qualities that determine the likelihood of exposure to a certain potentially hazardous entity.

Exposure to a certain synthetic-biological structure or system could be enhanced by that structure's ability to evolve and proliferate, by the absence of natural enemies, by the structure's increased persistence due to the formation of spore-like entities and by the structure's active and passive mobility, since exposure likelihood is particularly dependent on the quantity (and form) of a substance or synthetic-biological structure in a given system. With regard to synthetic-biological structures, proliferation seems to be the most relevant quality for increasing exposure and the emergence of risk. The rate of multiplication of biological entities is important, because it involves increased uncertainty about potential consequences. Proliferation can promote population growth up to ubiquitous occurrence. In their frequently cited publication on synthetic biology from 2006, Tucker and Zilinskas already identified self-replication and evolution-based development as qualities that justify a classification of corresponding organisms in a risk category different from toxic chemicals or radioactive material (cf. Tucker and Zilinskas 2006, p.31).

Besides exposure, specific *hazardous qualities* also contribute to potential risks (in an application context). The possible hazardous qualities of synthetic-biological structures are quite numerous. They include toxic effects of infectious microbes, viruses and phages, and molecules that lead to tissue damage or impaired functionality. Adverse toxic, ecological, global, and socio-technical effects could also be caused by interference with gene regulation, food competition with natural organisms, detrimental degradation of valuable or essential materials of the natural (e.g., lignocellulose) or built (e.g., metals, concrete) environment, and interference with bio-geo-chemical processes (e.g., nitrogen fixation, pH regulation). These hazardous qualities become effective if the natural or artificial environment (or in a legal sense: legally protected goods) are exposed to an altered or synthetic organism or to substances produced by those entities. All these detrimental effects are caused largely by external influences like displacement, degradation, infection, and colonization that leave the genetic integrity of natural organisms untouched. Moreover, adverse effects could also arise from genetic transfer and integration of an altered or synthetic transgene into a non-target organism. The transfer of genetic information bears a hazard potential if the encoded peptides, proteins or functional RNAs have toxic or pleiotropic effects on non-target organisms (cf. Mudgal et al. 2013, pp.41f.)[2].

---

2   See also Saxena and Stotzky (2001) regarding the higher lignin content in transgenic herbicide-tolerant BT corn plants.

In order to prevent an impact – or more precisely: adverse effects on the animate or inanimate environment – exposure and genetic transfers have to be prevented or at least minimized. To avoid exposure, containment is one possible solution. Here we can distinguish between physical containment, meaning that organisms are spatially separated by appropriate containers or tanks, and biological containment, where interference is prevented by orthogonality, which avoids interactions between synthetic and natural organisms at least on the genetic level (Marliere 2009; Schmidt and Lorenzo 2012). Due to the fact that present achievements in orthogonal biosystems are still far from being competent enough to establish the basis of an orthogonal organism (Moe-Behrens et al. 2013; Wright et al. 2013), effective biological containment will not represent an adequate option for critical applications of synthetic biology within the foreseeable future. And even if orthogonal systems would be available, one major disadvantage remains: potential effects can be caused by biochemical interactions on a metabolic level or by the displacement of natural organisms due to the proliferation of orthogonal entities. Therefore, we will consider only physical containment in the following classification. Two types of physical containment can be distinguished: a) closed systems and b) open systems in which contact to the environment is accepted. In contrast to the qualities of biological entities, which determine exposure and hazard characteristics, physical containment represents a quality of the operating conditions, hence the application context.

In short, some important qualities determining a hazard and/or exposure potential could be summarized by the terms containment, proliferation, and genetic transfer.[3] The difficulties encountered with respect to risk assessment and potential impact, and accordingly the probability of global and irreversible effects, increase along with the potential of proliferation, genetic transfer, and environmental release. Aside from a (rather hypothetical) complete physical isolation, the transfer of genetic information is hard to prevent. Wright et al. (2013) mention passive transfer of genetic material by transformation, meaning that DNA of dead cells could be taken up by other organisms, a mechanism that is even possible if the transgenic organism is equipped with a safety mechanism ensuring self-destruction: "Thus, even GMMs programmed to 'self-destruct' pose an environmental risk, as their genes can potentially be scavenged by other cells after they have died." (Wright et al. 2013, p.1222) The bare presence of synthetic (or altered) genetic information as the prerequisite for hazardous genetic transfer is therefore considered in a schematized illustration of major risk-relevant qualities (cf. Fig. 1).

---

3   Cp. the relevant criteria for biosafety in Wright et al. (2013).

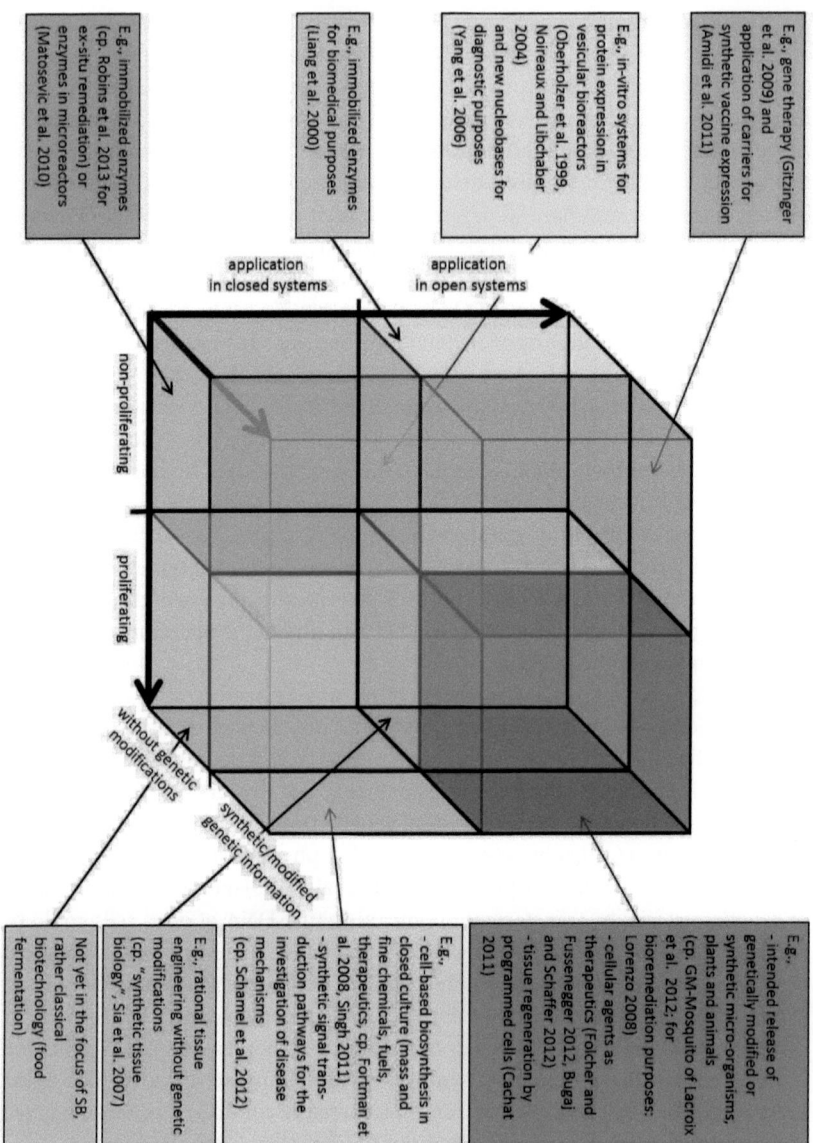

Fig. 1 Major risk-relevant qualities for applications in synthetic biology and examples of corresponding approaches (cyan to red ratio: cyan = less concern, red = most reasonable concern)

With regard to planned or already implemented applications of synthetic biology, at least one example is available for every section in the scheme of risk-relevant qualities – with the exception of genetically unmodified organisms (Fig. 1). These are more relevant for classical biotechnology and only touch on the creative sphere of synthetic biology in the case of tissue regeneration with the help of protein-based scaffolds. A number of techniques are at hand, especially for applications using organisms that are able to proliferate. However, the fact that the risks of those applications have been the subject of debate for several years demonstrates the existence of public awareness in this regard. Less spectacular but nonetheless promising applications might fall behind in the course of academic and political discussions about the use of synthetic biology in critical application contexts.

Consequently, the second part of this paper will further investigate selected approaches of the two fields of application: biological materials and energy generation, which might already offer interesting options for the use of synthetic biology. Moreover, these two fields have only a low potential of risk and indirect adverse impacts, respectively. Apart from risk potential, the selected examples are advantageous because of their potential in terms of resource consumption, energy generation, and the use of functionalities that are particularly linked to biological systems. The first field of application especially, biological materials, is characterized by unique functionalities that generate outstanding qualities. Due to biological processes of self-organization, complex hierarchical structures can be created whose application is of relatively low risk if separated from their production in closed cultures. Energy generation as the second field of application is a frequently discussed topic in synthetic biology. But the common strategy of producing biofuels from biomass with the help of microorganisms will be disregarded in this paper, because here we wish to focus on innovative concepts of energy generation that are less feedstock-intensive.

4 Biological materials

From an engineering perspective, nature has developed a broad variety of interesting solutions with quite a limited number of basic elements (Ashby et al. 1995; Dunlop and Fratzl 2012). Biological materials have numerous advantageous and often contrasting properties such as, for example, the combination of high elasticity and fracture strength, self-healing capability, and adaptive and multifunctional features (Fratzl and Barth 2009). These qualities are, for the most part, the result of certain combinations and specific structures at different hierarchical levels from molecules to tissues and organs. An important benefit of biological materials consists in their biodegradability[4]. For several years the imitation or synthesis of biological materials has been an ambitious aim of materials science and biomimetics. Another advantage lies in the manufacturing techniques. Technical structures are in many cases still produced in a waste-intensive, top-down approach by the stepwise *removal* of substance to obtain the intended shape or structure. The top-down approach was particularly unsuccessful in the synthesis of small, hierarchically structured materials because of its technical limits (Vincent 2008; Dunlop and Fratzl 2012). For that reason, research tended to address attempts to achieve functional morphology and investigate the mechanisms of self-organization. In contrast to the top-down approach, nature uses a bottom-up strategy where materials "grow" by processes of self-organization and differentiation (Fratzl and Barth 2009). Finally, artificial materials and systems usually fail to compete successfully with the adaptive capacity and variability of materials synthesized by organisms (Fratzl and Barth 2009). Therefore, new approaches towards the production of biological materials have to be investigated. Synthetic biology will most probably be an important discipline in the development of new biomimetic materials and structures.

4.1 Spider silk

Spider silk is a classic example of a biological material. As a protein fiber it combines high elasticity and tensile strength, currently unrivaled by any technical polymer. This is due to its characteristic hierarchical structure (Whitesides and Wong 2006): dragline silk (known for its outstanding properties) contains a number of small fibrillar structures, which themselves contain small ß-sheet crystalline protein regions separated by amorphous linking regions. The alternating order of the protein regions provides the above mentioned extraordinary qualities (Eisoldt et al. 2011). The modular composition of repetitive protein sequences regulates structural as well as functional qualities of the strand (Hayashi 2000; Brooks et al. 2008). Additionally, synthetic spider silk would be a biocompatible and biodegradable material, and it has been an important subject in the field of biological materials research for several years (Porter and Vollrath 2009; Eisoldt et al. 2011).

Despite knowledge of the molecular constitution of spider silk, a fiber with comparable qualities could not be synthesized up to now (Sponner et al. 2007). It turns out that the hierarchical structure of the fiber, in particular, plays a major

---

4 For a comprehensive presentation of functions of biological materials see the review of Liu and Jiang (2011).

role (Tarakanova and Buehler 2012) and therefore the main emphasis of research is on protein folding in the course of the spinning process (Vollrath 2000; Vollrath and Knight 2001). Attempts to farm spiders were quite difficult and less productive because of their territorial, cannibalistic behavior. Biotechnological approaches have so far suffered from the fact that it has proved difficult to transfer the ability to produce spider silk proteins into other host organisms (e. g., bacteria, mammalian cells, plants or even silkworms). In comparison to natural spider silk, only small proteins in low amounts could be expressed (Xia et al. 2010; Chung et al. 2012). Moreover, purification after cellular disruption and the spinning process of the protein fibers represent two additional obstacles. But purification and concentration of the respective proteins have now been improved by harnessing a secretion mechanism from *Salmonella* for the export of expressed proteins (Widmaier et al. 2009). This approach demonstrates the potential of synthetic biology for the synthesis of biological materials (Chung et al. 2012). Nevertheless, the spinning process remains difficult because, for a successful assembly, the proteins must be in solution, which has been hard to achieve until now (ibid.). The targeted modification of certain amino acids might serve to alter the properties of the spider silk protein accordingly (as in the case of biological adhesives described in the following section). The most promising applications are technical textiles, clothing, and medicine.

4.2 Biological adhesives

With the help of adhesive fibers, a variety of mussels are able to adhere to different substrate surfaces under water, even in difficult environmental conditions. These strong adhesive fibers called byssus are synthesized in the mussel's byssus gland. The biological adhesion of *Mytilus edulis*, the blue mussel, has been best characterized, but other species such as *Mytilus californianus* are also being investigated (Wiegemann 2005; Nicklisch and Waite 2012). Common biological adhesives consist of a combination of proteins, polysaccharides, polyphenoles, and lipids (Wiegemann 2005). A byssus filament contains at least ten different proteins, which form four different segments, each with a distinct constitution according to its specific tasks. The byssal thread is composed of elastin and collagen (Silverman and Roberto 2007). Whereas collagen gives tensile strength, elastin absorbs shocks. The byssal plaque (or 'foot' of the byssal thread), which is responsible for adhesion to the substrate, consists of six different proteins, also known as "mussel foot proteins" (Mfp) (Grunwald et al. 2009; Nicklisch and Waite 2012). Besides their amino acid sequence, Mfps differ especially in their amount of one specific amino acid: L-3,4-dihydroxyphenylalanin (DOPA). DOPA is not a common amino acid, but mussels have the ability to convert tyrosine into DOPA by hydroxylation.

In the production of byssus fiber as a biological adhesive the integration of DOPA is difficult to achieve in other host systems. This problem may be solved by synthetic biology. For the purpose of incorporating a synthetic amino acid, the research group of Nediljko Budisa modified proline with fluorine and tested the flexibility of the endogenous bacterial translation machinery (Larregola et al. 2012). They showed that the translation was successful despite the fact that the proline had been fluorinated. The introduced alteration may have a beneficial effect on the Mfp because fluorine is known for its stabilizing effect in proteins; this is currently being further investigated (Larregola et al. 2012).

Applications of mussel-derived adhesives are primarily found in the medical field, especially in dentistry, because mussel adhesives are biocompatible and have the ability to stick to various substrates in aqueous solutions (e.g., glass, plastic, metal, wood, bone tissue, and other biological tissues and surfaces) (Silverman and Roberto 2007). In addition, DOPA could, because of its variable substrate affinity, be used – among other applications – for biomarkers (Brubaker and Messersmith 2012).

4.3 Synthetic Nacre

Natural nacre is a composite material consisting of materials with different degrees of hardness. Alternating layers of calcium carbonate in the form of aragonite platelets are separated by a soft organic layer of chitin and proteins. The organic matrix acts as a template that steers the specific (self-organized) crystallization of the platelets. Due to the characteristic stacking order of the hard aragonite and a flexible matrix in between, nacre has a 3,000 times greater hardness than the sum of its components (Barthelat and Zhu 2011). Thus, only very strong mechanical forces can lead to fissures or crack the nacre. Because of this hardness *in combination with* its toughness and resistance to fracture, nacre (and its structure in general) serves as an interesting inspiration for high-tech materials (Sun and Bhushan 2012) and has been a field of investigation for several years. But until recently no comparable emergent strengthening effects could be achieved by the combination of separate materials (Barthelat and Zhu 2011). In 2011, Laaksonen et al. successfully constructed a nacre-related composite using nanocellulose, graphene, and a diblock-protein. The protein component was genetically modified to allow the incorporation of a hydrophobin block to bind graphene and a cellulose-binding block for the linkage to nanocellulose (Laaksonen et al. 2011). The properties of the resulting composite are comparable to natural nacre.

4.4 Challenges to biological materials

The field of synthetic biological materials currently faces two important obstacles: the formation of a desired hierarchical structure by self-organization processes, and suitable host organisms for the expression of biomolecules. Previous host organisms proved to be limited. For example, in the case of spider silk the chosen host was not able to produce proteins the size of natural silk proteins at the intracellular level. This limitation could admittedly be circumvented by the implementation of a secretion system, but nevertheless a successful approach for the synthesis of filaments with a defined molecular structure is still lacking. Therefore, synthetic silk fails to reach the material qualities of natural spider silk (Keerl and Scheibel 2012).

Directed (template-induced) self-organization of biological materials across multiple scales is still challenging. On the one hand, there is insufficient knowledge regarding changes of properties during shifts from the quantum scale to the macroscale (Buehler 2010). On the other hand, control of directed self-organization must be ensured in a bottom-up approach. The intended material properties at the macroscale are determined by subscale levels. For instance, in the case of spider silk the sequence of amino acids represents the lowest level which determines all subsequent scales (Tarakanova and Buehler 2012). A corresponding approach of synthetic biology consists in the 'periodic system' of self-organizing functional proteins ('tectons') developed by Woolfson et al. (Bromley et al. 2008; Moutevelis and Woolfson 2009). If processes of biological self-organization and their dependencies across multiple scales were sufficiently understood, material properties could be influenced more specifically. Some initial attempts in molecular modeling have been undertaken, but precise methods for an analysis on the molecular level of in-silico models are still lacking (Tarakanova and Buehler 2012). However, a paradigm shift seems to be necessary considering the difference between current claims of control and determination in engineering on the one hand, and incomparable degrees of freedom and variability for processes of self-organization on the other.

Besides biological materials, a second quite popular application field of synthetic biology will be discussed here: the generation of useful energy (exergy). In contrast to well-known and much-debated approaches, a number of techniques that avoid the risks of open cultivation or indirect adverse effects due to the consumption of arable land and biomass are presented here.

5 Energy applications

Envisaged applications of synthetic biology in the context of energy generation are mainly based on the same principle as classical biotechnological processes: biomass, usually of plant origin, is transformed by microorganisms into energy-rich combustibles. But for that purpose plant biomass has to be provided, for which areas of arable land, water, fertilizers (associated especially with the emission of gases having high global warming potential) and also energy are required. Thus, synthetic biology is confronted with the same conflicts as found in connection with large-scale biomass consumption in previous biotechnological processes. The competition between agricultural land used for food and land used for energy crops is of particular importance. Moreover, a number of other ecological and social consequences, which generally accompany large-scale cultivation of biomass, are also relevant to synthetic-biological approaches for energy harvesting (FoE 2010; ETC 2010; Das et al. 2010). Quite a few investigations have revealed that the cultivation of biomass for energy generation may need decades, or even centuries, from its initial introduction until it may need decades, or even centuries, from its initial introduction until it fulfills one of its main purposes, namely the reduction of greenhouse-gas emission (Searchinger et al. 2008; Fargione et al. 2008).

The focus of this review is therefore on approaches that could – at least partially – help us avoid some of these conflicts and risks.

5.1 Chemoautotrophic and (bio-) electrosynthetic approaches

Certain microorganisms known as chemoautotrophs are naturally capable of synthesizing organic compounds from carbon dioxide and water without using light (photons) as an energy source (this also applies to plants, algae, and other microorganisms). Instead, chemoautotrophs use inorganic substances to acquire energy (electrons) needed in carbon fixation. Independence from light sources may be an important advantage for biotechnological processes because such organisms can be cultivated in compact, closed reactors independent of solar light intensity at the production sites. Together with their autotrophic character, namely autonomy from biomass as a substrate which minimizes agricultural conflicts, this property makes them an interesting alternative for the production of biofuels. Finally, chemoautotrophs could help to establish long-term storage of energy from renewable energy plants by synthesizing liquid fuels of high energy density from less advantageous hydrogen produced by electrolysis. Because of all these possible benefits, chemoautotrophs are the subject of (synthetic) biological, biochemical, and biotechnological research (Fast and Papoutsakis 2012; Hawkins et al. 2013; Li and Liao 2013). Some examples include the bacterial species *Ralstonia eutropha*, *Ralstonia europea*, some *Clostridia* species as well as the Archaea *Pyrococcus furiosus* and *Metallosphaera sedula* (Lovley and Nevin 2013; Hawkins et al. 2013).

Two strategies of genetic engineering and metabolic engineering, respectively, are applied here: on the one hand, biofuel-producing pathways are implemented in autotrophic organisms; on the other hand, the aim is to couple carbon fixation with native pathways of heterotrophic organisms used for biofuel synthesis (Hawkins et al. 2013). Some microorganisms are even able to directly accept electrons for carbon fixation. Hence, certain approaches try to exploit this quality for an electrically driven synthesis of biomolecules (electrosynthesis) in order to – for example – solve the storage problem of renewable energy from wind and solar power plants (Rabaey et al. 2011; Lovley and Nevin 2013). In an optimal scenario for energy transformation, autotrophic microorganisms would reduce carbon dioxide to organic compounds with the help of water and externally provided electrons. These compounds could be stored as fuels and used only when required. In this way, carbon would be sequestered, thereby avoiding carbon dioxide emissions which increase global warming. Furthermore, no arable land would be required for this approach because such microorganisms would need relatively few or no other organic materials (from biomass). For that reason, conflicts with food production could be minimized. Alternatively, potential processes could function in a partially or completely heterotrophic modus, optimized by the supply of electrons (e.g., by influencing product composition). In terms of biological and biochemical aspects, as well as regarding the characteristics of processes and materials, electrosynthesis is closely related to the approach of bio-electricity (see below), with the effect that both attempts even benefit from each other (Rabaey et al. 2011).

In light of the several steps of energy conversion, electrosynthesis would appear to be an inefficient process at first glance. But in fact, under certain conditions the electrosynthesis approach could gain a higher total efficiency in converting solar to chemically bound energy than systems relying on photosynthesis. This advantage is primarily due to the significantly higher efficiency of photovoltaics in comparison to natural photosynthesis of plants and algae. Though biomass from the latter is seldom used as fuel because it requires further steps connected with losses in mass and energy (Lovley and Nevin 2013). Species currently being investigated because of their potential for electrosynthesis include *Morella thermoacetica* and some *Sporomusa* and *Clostridium* species (Lovley and Nevin 2013).

5.2 Bio-electricity

The vast majority of research and development pertaining to energy generation from biomass focuses on fuels that are either used in combustion engines or fuel cells, as in the case of hydrogen. But there have also been some attempts to obtain electricity directly from microorganisms (Kiely et al. 2011) or from cell-free systems (Cooney et al. 2008). In analogy to hydrogen-powered fuel cells, a 'microbial fuel cell' would be a source of electricity.

At present, a number of bacterial species that deliver electrons naturally (i.e., without any genetic modifications) are known. They include *Escherichia Coli* and *Pelobacter propionicus* and also bacteria of the genera *Geobacter, Shewanella, Azoarcus, Desulfuromonas, Thauera* and *Pseudomonas* (Yong et al. 2011; Kiely et al. 2011). The functional principle and parameters of electrical performance can be optimized by genetic engineering (Yong et al. 2011). Electron-donating microorganisms have a broad range of substrates and could therefore also grow in wastewater (Kiely et al. 2011). Electron-donating degradation processes are quite complex and include multiple stages in which a variety of different bacterial species and genera is synergistically involved (Kiely et al. 2011). Many of the mechanisms involved are still unexplored.

In the case of cell-free systems, investigations focus on the optimization of natural pathways and the development of artificial enzymatic combinations. Within the field of 'enzyme engineering', the catalytically active proteins involved are optimized as well (Guterl and Sieber 2013). Furthermore, a number of issues regarding process engineering, materials for electrodes, and the design of the reactors used are currently being explored (Cooney et al. 2008).

Today, the field is largely dominated by basic research trying to uncover mechanisms of exogenic electron donation. Without at least a basic knowledge, it is hard to find starting points for effective intervention by synthetic-biological manipulation or even a de-novo design of structures and organisms. This deficit may be the reason for putting the main emphasis on optimizing process parameters such as the structure and material of electrodes, or the composition of the applied microbial communities, where the underlying bio(electro)chemical processes are treated as a 'black box'.

6 Conclusion

The specific approach of synthetic biology offers a number of interesting solutions for different application fields. However, a more comprehensive analysis of perceived advantageous techniques is required to assess their feasibility, and to differentiate between really sound and sustainable solutions on the one hand, and technologies that might lead to already known or even new detrimental effects on the other. As we have endeavored to show, the analysis of inherent qualities of new synthetic-biological processes and constructs is essential for an early evaluation of chances and risks, especially with regard to major risk-relevant qualities and operating conditions. The latter include, with regard to exposure: potential for proliferation, genetic transfer or insufficient containment.

Besides synthesis as a complementation for empirical forms of biological analysis, the implementation of rational engineering principles into biology – as well as biotechnology – represents a substantial extension of basic and applied approaches of the field. Nevertheless, it is still questionable how far engineering principles and their claims of comprehensive control can be transferred to biological systems. Synthetic biology may make a significant contribution to the fields of applied biology, provided it can be adapted to the requirements of industrial ecology (i.e., sustainable integration of the energy and material flows of the technosphere into the energy and material flows of the ecosphere). One important step in this direction could be the development of biodegradable high-technology materials with outstanding properties by biomimesis. This would close the gaps in material cycles by using waste materials, instead of consuming large amounts of land and biomass in competition with food production and nature conservation. At the same time, while we improve the efficiency of biological constructs through a rational and comprehensive approach, risks have to be reduced as well in a planned, systematic approach.

In this context we have shown that a) the field of biological materials especially may harness the unique quality of biological systems to produce hierarchically structured matter, and b) biological forms of energy generation may overcome the extensive consumption of biomass or arable land by drawing on certain, as yet uncommon, inherent processes of microorganisms.

Applications of synthetic biology, which benefit from the unique features of biological systems, may be decoupled from already well-known hazard potentials like uncontrolled distribution and proliferation due to their use in open systems. Moreover, there are already a number of valuable technologies at hand that make use of non-living, cell-free approaches (Doktycz and Simpson 2007; Puri et al. 2009; Shin and Noireaux 2012; Nourian et al. 2012; Kim and Winfree 2011; Robins et al. 2013), thus eliminating the risks that would otherwise arise from living (i.e., evolving and proliferating) systems.

Finally, we conclude that there could be several opportunities for a sustainable use of the technological functionalities provided by synthetic biology. It is important that they be chosen carefully and, most of all, that they avoid the paths of previous attempts that proved to have adverse side effects.

Acknowledgement

This work has been funded by the German Ministry of Education and Research (BMBF) as part of the study "Technology Assessment of Synthetic Biology" under project code 16I1611. We are grateful for the opportunity to conduct a research project on synthetic biology with particular focus on its scientific and technological character, achievements, and functionalities as well as consequences thereof.

References

Amidi, M., Raad, M., Crommelin, D.A., Hennink, W., & Mastrobattista, E. (2011). Antigen-expressing immunostimulatory liposomes as a genetically programmable synthetic vaccine. *Systems and Synthetic Biology, 5*(1-2), 21-31. doi: 10.1007/s11693-010-9066-z.

Ashby, M.F., Gibson, L.J., Wegst, U., & Olive, R. (1995). The Mechanical Properties of Natural Materials. I. Material Property Charts. *Proceedings of the Royal Society. Mathematical and Physical Sciences, 450*(1938), 123-140.

Barthelat, F., & Zhu, D. (2011). A novel biomimetic material duplicating the structure and mechanics of natural nacre. *Journal of Materials Research, 26*(10), 1203-1215. doi: 10.1557/jmr.2011.65.

Bromley, E.H.C., Channon, K., Moutevelis, E., & Woolfson, D.N. (2008). Peptide and protein building blocks for synthetic biology: From programming biomolecules to self-organized biomolecular systems. *ACS Chemical Biology, 3*(1), 38-50. doi: http://dx.doi.org/10.1021/cb700249v.

Brooks, A.E., Stricker, S.M., Joshi, S.B., Kamerzell, T.J., Middaugh, C.R., & Lewis, R.V. (2008). Properties of synthetic spider silk fibers based on Argiope aurantia MaSp2. *Biomacromolecules, 9*(6), 1506-1510. doi: 10.1021/bm701124p.

Brubaker, C.E., & Messersmith, P.B. (2012). The present and future of biologically inspired adhesive interfaces and materials. *Langmuir, 28*(4), 2200-2205. doi: 10.1021/la300044v.

Buehler, M.J. (2010). Multiscale Mechanics of Biological and Biologically Inspired Materials and Structures. *Acta Mechanica Solida Sinica, 23*(6), 471-483.

Bugaj, L.J., & Schaffer, D.V. (2012). Bringing Next-Generation Therapeutics to the Clinic through Synthetic Biology. *Current Opinion in Chemical Biology, 16*(3-4), 355-361. doi: 10.1016/j.cbpa.2012.04.009.

Cachat, E., & Davies, J.A. (2011). Application of Synthetic Biology to Regenerative Medicine. *Journal of Bioengineering and Biomedical Sciences, S2*, 1-9. doi: 10.4172/2155-9538.s2-003.

Canton, B., Labno, A., & Endy, D. (2008). Refinement and standardization of synthetic biological parts and devices. *Nature Biotechnology, 26*(7), 787-793. doi: 10.1038/nbt1413.

Chung, H., Kim, T.Y., & Lee, S.Y. (2012). Recent advances in production of recombinant spider silk proteins. *Current Opinion in Biotechnology, 23*(6), 957-964. doi: 10.1016/j.copbio.2012.03.013.

Cooney, M.J., Svoboda, V., Lau, C., Martin, G., & Minteer, S.D. (2008). Enzyme catalysed biofuel cells. *Energy & Environmental Science, 1*(3), 320-337.

Das, S., Priess, J.A., & Schweitzer, C. (2010). Biofuel Options for India-Perspectives on Land Availability, Land Management and Land-Use Change. *Journal of Biobased Materials and Bioenergy, 4*(3), 243-255. doi: 10.1166/jbmb.2010.1089.

DeWall, M.T., & Cheng, D.W. (2011). The minimal genome: a metabolic and environmental comparison. *Briefings in Functional Genomics, 10*(5), 312-315. doi: 10.1093/bfgp/elr030.

Dietz, S., & Panke, S. (2010). Microbial systems engineering: first successes and the way ahead. *Bioessays, 32*(4), 356-362. doi: 10.1002/bies.200900174.

Doktycz, M.J., & Simpson, M.L. (2007). Nano-enabled synthetic biology. *Molecular Systems Biology, 3*, 125. doi: 10.1038/msb4100165.

Dunlop, J.W.C., & Fratzl, P. (2012). Multilevel architectures in natural materials. *Scripta Materialia, 68*(1), 8-12. doi: 10.1016/j.scriptamat.2012.05.045.

Dymond, J.S., Richardson, S.M., Coombes, C.E., Babatz, T., Muller, H., Annaluru, N., . . . Boeke, J.D. (2011). Synthetic chromosome arms function in yeast and generate phenotypic diversity by design. *Nature, 477*(7365), 471-476.

Eisoldt, L., Smith, A., & Scheibel, T. (2011). Decoding the secrets of spider silk. *Materials Today, 14*(3), 80-86. doi: 10.1016/S1369-7021(11)70057-8.

Eldar, A. & Elowitz, M.B. (2010). Functional roles for noise in genetic circuits. *Nature, 467*(7312), 167-173.

ETC (2010). ETC Group. *The New Biomassters. Synthetic Biology and the Next Assault on Biodiversity and Livelihoods.* http://www.etcgroup.org/sites/www.etcgroup.org/files/biomassters_27feb2011.pdf. Accessed: 24 March 2014.

Evonik (2013). Synthesegas schmeckt Bakterien. Press release of 4 December 2013. Evonik Industries AG. http://corporate.evonik.de/de/presse/suche/pages/news-details.aspx-?newsid=40323. Accessed: 21 May 2015.

Fargione, J., Hill, J., Tilman, D., Polasky, S., & Hawthorne, P. (2008). Land clearing and the biofuel carbon debt. *Science, 319*(5867), 1235-1238. doi: 10.1126/science.1152747.

Fast, A.G., & Papoutsakis, E.T. (2012). Stoichiometric and energetic analyses of non-photosynthetic CO_2-fixation pathways to support synthetic biology strategies for production of fuels and chemicals. *Current Opinion in Chemical Engineering, 1*(4), 380-395. doi: 10.1016/j.coche.2012.07.005.

FoE (2010). Friends of the Earth. *Synthetic Solutions to the Climate Crisis: The Dangers of Synthetic Biology for Biofuels Production.* Report. http://libcloud.s3.amazonaws.com/93/59/9/529/1/SynBio-Biofuels_Report_Web.pdf. Accessed: 25 March 2014.

Folcher, M., & Fussenegger, M. (2012). Synthetic Biology Advancing Clinical Applications. *Current Opinion in Chemical Biology, 16*(3-4), 345-354. doi: 10.1016/j.cbpa.2012.06.008.
Fortman, J.L., Chhabra, S., Mukhopadhyay, A., Chou, H., Lee, T.S., Steen, E., & Keasling, J.D. (2008). Biofuel alternatives to ethanol: pumping the microbial well. *Trends in biotechnology, 26*(7), 375-381.
Fratzl, P., & Barth, F.G. (2009). Biomaterial systems for mechanosensing and actuation. *Nature, 462*(7272), 442-448. doi: 10.1038/nature08603.
Gitzinger, M., Kemmer, C., El-Baba, M. D., Weber, W., & Fussenegger, M. (2009). Controlling transgene expression in subcutaneous implants using a skin lotion containing the apple metabolite phloretin. *Proceedings of the National Academy of Sciences, 106*(26), 10638-10643. doi: 10.1073/pnas.0901501106.
Grunwald, I., Rischka, K., Kast, S.M., Scheibel, T., & Bargel, H. (2009). Mimicking biopolymers on a molecular scale: nano(bio)technology based on engineered proteins. *Philosophical transactions. Series A, Mathematical, physical, and engineering sciences, 367*(1894), 1727-1747. doi: 10.1098/rsta.2009.0012.
Guterl, J.-K., & Sieber, V. (2013). Biosynthesis "debugged": Novel bioproduction strategies. *Engineering in Life Sciences, 13*(1), 4-18. doi: 10.1002/elsc.201100231.
Hawkins, A.S., McTernan, P.M., Lian, H., Kelly, R.M. &, Adams, M.W.W. (2013). Biological conversion of carbon dioxide and hydrogen into liquid fuels and industrial chemicals. *Current Opinion in Biotechnology, 24*(3), 376-384. doi: 10.1016/j.copbio.2013.02.017.
Hayashi, C.Y. (2000). Molecular Architecture and Evolution of a Modular Spider Silk Protein Gene. *Science, 287*(5457), 1477-1479. doi: 10.1126/science.287.5457.1477.
Kastenhofer, K., & Schmidt, J.C. (2011). On Intervention, Construction and Creation: Power and Knowledge in Technoscience and Late-Modern Technology. In: T.B. Zülsdorf, C. Coenen, A. Ferrari, U. Fiedeler, C. Milburn, M. Wienroth (eds.), Quantum *Engagements: Social Reflections of Nanoscience and Emerging Technologies* (pp.177-194). Heidelberg: Akademische Verlagsgesellschaft.
Keerl, D., & Scheibel, T. (2012). Characterization of natural and biomimetic spider silk fibers. *Bioinspired, Biomimetic and Nanobiomaterials, 1*(2), 83-94. doi: 10.1680/bbn.11.00016.
Khalil, A.S., & Collins, J.J. (2010). Synthetic biology: applications come of age. *Nature Reviews Genetics, 11*(5), 367-379.
Kiely, P.D., Regan, J.M., & Logan, B.E. (2011). The Electric Picnic: Synergistic Requirements for Exoelectrogenic Microbial Communities. *Current Opinion in Biotechnology, 22*(3), 378-385. doi: 10.1016/j.copbio.2011.03.003.
Kim, J., & Winfree, E. (2011). Synthetic in vitro transcriptional oscillators. *Molecular Systems Biology, 7*(465), 465-465. doi: 10.1038/msb.2010.119.
Kittleson, J.T., Wu, G.C., & Anderson, J.C. (2012). Successes and failures in modular genetic engineering. *Current Opinion in Chemical Biology, 16*(3-4), 329-336. doi: 10.1016/j.cbpa.2012.06.009.
Laaksonen, P., Walther, A., Malho, J.-M., Kainlauri, M., Ikkala, O., & Linder, M. B. (2011). Genetic Engineering of Biomimetic Nanocomposites: Diblock Proteins, Graphene, and Nanofibrillated Cellulose. *Angewandte Chemie International Edition, 50*, 8688-8691. doi: 10.1002/anie.201102973.
Lacroix, R., McKemey, A.R., Raduan, N., Wee, L.K., Ming, W.H., Ney, T.G., ... Murad, S. (2012). Open field release of genetically engineered sterile male Aedes aegypti in Malaysia. *PLOS ONE, 7*(8), e42771.

LaMonica, M. (2014). Why the Promise of Cheap Fuel from Super Bugs Fell Short. *MIT Technology Review*, 5 February 2014. http://www.technologyreview.com/news/524011/why-the-promise-of-cheap-fuel-from-super-bugs-fell-short/. Accessed: 8 July 2014.

Larregola, M., Moore, S., & Budisa, N. (2012). Congeneric bio-adhesive mussel foot proteins designed by modified prolines revealed a chiral bias in unnatural translation. *Biochemical and Biophysical Research Communications, 421*(4), 646-650. doi: 10.1016/j.bbrc.2012.04.031.

Li, H., & Liao, J.C. (2013). Biological conversion of carbon dioxide to photosynthetic fuels and electrofuels. *Energy & Environmental Science, 6*(10), 2892-2899. doi: 10.1039/c3ee41847b.

Liang, J.F., Li, Y.T., & Yang, V.C. (2000). Biomedical Application of Immobilized Enzymes. *Journal of Pharmaceutical Sciences, 89*(8), 979-990.

Liu, K. & Jiang, L. (2011). Bio-inspired design of multiscale structures for function integration. *Nano Today, 6*(2), 155-175. doi: 10.1016/j.nantod.2011.02.002.

Lorenzo, V. d. (2008). Systems biology approaches to bioremediation. *Current Opinion in Biotechnology, 19*(6), 579-589. doi: 10.1016/j.copbio.2008.10.004.

Lovley, D.R., & Nevin, K.P. (2013). Electrobiocommodities: powering microbial production of fuels and commodity chemicals from carbon dioxide with electricity. *Current Opinion in Biotechnology, 24*(3), 385-390. doi: 10.1016/j.copbio.2013.02.012.

Marliere, P. (2009). The farther, the safer: a manifesto for securely navigating synthetic species away from the old living world. *Systems and Synthetic Biology, 3*, 77-84.

Matosevic, S., Lye, G.J., & Baganz, F. (2010). Design and characterization of a prototype enzyme microreactor: quantification of immobilized transketolase kinetics. *Biotechnology Progress, 26*(1), 118-126. doi: 10.1002/btpr.319.

Moe-Behrens, G.H., Davis, R., & Haynes, K.A. (2013). Preparing synthetic biology for the world. *Frontiers in Microbiology, 4*, 1-10. doi: 10.3389/fmicb.2013.00005.

Moutevelis, E., & Woolfson, D.N. (2009). A periodic table of coiled-coil protein structures. *Journal of Molecular Biology, 385*(3), 726-732. doi: 10.1016/j.jmb.2008.11.028.

Mudgal, S., Toni, A. d., Tostivint, C., Hokkanen, H., & Chandler, D. (2013). *Scientific support, literature review and data collection and analysis for risk assessment on microbial organisms used as active substance in plant protection products –Lot 1 Environmental Risk characterisation*. EFSA supporting publications 2013: EN-518. http://www.efsa.europa.eu/en/efsajournal/doc/518e.pdf. Accessed: 21 May 2015.

Mutalik, V.K., Guimaraes, J.C., Cambray, G., Lam, C., Christoffersen, M.J., Mai, Q.A., .. . Endy, D. (2013). Precise and reliable gene expression via standard transcription and translation initiation elements. *Nature methods, 10*(4), 354-360.

Nicklisch, S.C., & Waite, J.H. (2012). Mini-review: the role of redox in Dopa-mediated marine adhesion. *Biofouling, 28*(8), 865-877. doi: 10.1080/08927014.2012.719023.

Noireaux, V., & Libchaber, A. (2004). A vesicle bioreactor as a step toward an artificial cell assembly. *Proceedings of the National Academy of Sciences of the United States of America, 101*(51), 17669-17674. doi: 10.1073/pnas.0408236101.

Nordmann, A. (2011). Science in the Context of Technology. *Boston Studies in the Philosophy and History of Science, 274*(6), 467-482.

Nourian, Z., Roelofsen, W., & Danelon, C. (2012). Triggered gene expression in fed-vesicle microreactors with a multifunctional membrane. *Angewandte Chemie International Edition, 51*(13), 3114-3118. doi: 10.1002/anie.201107123.

Oberholzer, T., Nierhaus, K.H., & Luisi, P.L. (1999). Protein Expression in Liposomes. *Biochemical and Biophysical Research Communications, 261*(2), 238-241.

Paschen, H., & Petermann, T. (1992). Technikfolgen-Abschätzung. Ein strategisches Rahmenkonzept für die Analyse und Bewertung von Techniken. In: T. Petermann (ed.), *Technikfolgen-Abschätzung als Technikforschung und Politikberatung* (pp.19-42). Frankfurt/Main: Campus.

Porter, D., & Vollrath, F. (2009). Silk as a Biomimetic Ideal for Structural Polymers. *Advanced Materials, 21*(4), 487-492. doi: 10.1002/adma.200801332.

Puri, A., Loomis, K., Smith, B., Lee, J.H., Yavlovich, A., Heldman, E., & Blumenthal, R. (2009). Lipid-based nanoparticles as pharmaceutical drug carriers: from concepts to clinic. *Critical Reviews in Therapeutic Drug Carrier Systems, 26*(6), 523-580.

Rabaey, K., Girguis, P. & Nielsen, L.K. (2011). Metabolic and Practical Considerations on Microbial Electrosynthesis. *Current Opinion in Biotechnology, 22*(3), 371-377. doi: 10.1016/j.copbio.2011.01.010.

Robins, K.J., Hooks, D.O., Rehm, B.H., & Ackerley, D.F. (2013). Escherichia coli NemA is an efficient chromate reductase that can be biologically immobilized to provide a cell free system for remediation of hexavalent chromium. *PLOS ONE, 8*(3), e59200. doi: 10.1371/journal.pone.0059200.

Saxena, D., & Stotzky, G. (2001). BT corn has a higher lignin content than non-BT corn. *American Journal of Botany, 88*(9), 1704-1706.

Schamel, W.W.A., & Reth, M. (2012). Synthetic immune signaling. *Current Opinion in Biotechnology, 23*(5), 780-784. doi: 10.1016/j.copbio.2012.01.010.

Schmidt, M., & Lorenzo, V. d. (2012). Synthetic constructs in/for the environment: managing the interplay between natural and engineered Biology. *FEBS Letters, 586*(15), 2199-2206. doi: 10.1016/j.febslet.2012.02.022.

Searchinger, T., Heimlich, R., Houghton, R.A., Dong, F., Elobeid, A., Fabiosa, J., ... Yu, T.H. (2008). Use of US croplands for biofuels increases greenhouse gases through emissions from land-use change. *Science, 319*(5867), 1238-1240.

Shin, J., & Noireaux, V. (2012). An E. coli cell-free expression toolbox: application to synthetic gene circuits and artificial cells. *ACS Synthetic Biology, 1*(1), 29-41. doi: 10.1021/sb200016s.

Sia, S.K., Gillette, B.M., & Yang, G.J. (2007). Synthetic tissue biology: Tissue engineering meets synthetic biology. *Birth Defects Research Part C: Embryo Today: Reviews, 81*(4), 354-361. doi: 10.1002/bdrc.20105.

Silverman, H.G. & Roberto, F.F. (2007). Understanding marine mussel adhesion. *Marine Biotechnology, 9*(6), 661-681. doi: 10.1007/s10126-007-9053-x.

Singh, R. (2011). Facts, Growth, and Opportunities in Industrial Biotechnology. *Organic Process Research & Development, 15*(1), 175-179.

Sponner, A., Vater, W., Monajembashi, S., Unger, E., Grosse, F., & Weisshart, K. (2007). Composition and Hierarchical Organisation of a Spider Silk. *PLOS ONE, 2*(10), e998. doi: 10.1371/journal.pone.0000998.

Stanton, B.C., Nielsen, A.A.K., Tamsir, A., Clancy, K., Peterson, T., & Voigt, C.A. (2013). Genomic mining of prokaryotic repressors for orthogonal logic gates. *Nature Chemical Biology, 10*, 99-105. doi: 10.1038/nchembio.1411.

Sun, J., & Bhushan, B. (2012). Hierarchical structure and mechanical properties of nacre: a review. *RSC Advances, 2*(20), 7617. doi: 10.1039/c2ra20218b.

Tarakanova, A., & Buehler, M.J. (2012). A Materiomics Approach to Spider Silk: Protein Molecules to Webs. *Jom, 64*(2), 214-225. doi: 10.1007/s11837-012-0250-3.

Tucker, J., & Zilinskas, R. (2006). The promise and perils of synthetic biology. *The New Atlantis, 12*(1), 25-45.

Vincent, J.F.V. (2008). Biomimetic Materials. *Journal of Materials Research, 23*(12), 3140-3147. doi: 10.1557/jmr.2008.0380.

Vollrath, F. (2000). Strength and structure of spiders' silks. *Journal of Biotechnology, 74*(2), 67-83.

Vollrath, F., & Knight, D.P. (2001). Liquid crystalline spinning of spider silk. *Nature, 410*(6828), 541-548. doi: 10.1038/35069000.

Wang, H.H., & Church, G.M. (2011). Multiplexed Genome Engineering and Genotyping Methods: Applications for Synthetic Biology and Metabolic Engineering. *Methods in Enzymology, 498*, 409-426.

Whitesides, G.M., & Wong, A.P. (2006). The Intersection of Biology and Materials Science. *MRS Bulletin, 31*(01), 19-27. doi: 10.1557/mrs2006.2.

Widmaier, D.M., Tullman-Ercek, D., Mirsky, E.A., Hill, R., Govindarajan, S., Minshull, J., & Voigt, C.A. (2009). Engineering the Salmonella type III secretion system to export spider silk monomers. *Molecular Systems Biology, 5*(1), 309. doi: 10.1038/msb.2009.62.

Wiegemann, M. (2005). Adhesion in blue mussels (Mytilus edulis) and barnacles (genus Balanus): Mechanisms and technical applications. *Aquatic Sciences, 67*(2), 166-176. doi: 10.1007/s00027-005-0758-5.

Wright, O., Stan, G.B., & Ellis, T. (2013). Building-in biosafety for synthetic biology. *Microbiology, 159*(Pt 7), 1221-1235. doi: 10.1099/mic.0.066308-0.

Xia, X.-X., Qian, Z.-G., Ki, C.S., Park, Y.H., Kaplan, D.L., & Lee, S.Y. (2010). Native-sized recombinant spider silk protein produced in metabolically engineered Escherichia coli results in a strong fiber. *Proceedings of the National Academy of Sciences of the United States of America, 107*(32), 14059-14063. doi: 10.1073/pnas.1003366107.

Yang, Z., Hutter, D., Sheng, P., Sismour, A.M., & Benner, S.A. (2006). Artificially expanded genetic information system: a new base pair with an alternative hydrogen bonding pattern. *Nucleic Acids Research, 34*(21), 6095-6101. doi: 10.1093/nar/gkl633.

Yeh, B.J., & Lim, W.A. (2007). Synthetic biology: lessons from the history of synthetic organic chemistry. *Nature Chemical Biology, 3*(9), 521-525.

Yong, Y.-C., Yu, Y.-Y., Li, C.-M., Zhong, J.-J., & Song, H. (2011). Bioelectricity Enhancement via Overexpression of Quorum Sensing System in *Pseudomonas aeruginosa*-Inoculated Microbial Fuel Cells. *Biosensors & Bioelectronics, 30*(1), 87-92. doi: 10.1016/j.bios.2011.08.032.

Synthetic biology's multiple dimensions of benefits and risks: implications for governance and policies

Harald König, Daniel Frank, Reinhard Heil and Christopher Coenen

1 Introduction

Synthetic biology (SB) does not constitute a strictly defined field, but may be best described as an engineering approach aimed at redesigning or newly constructing biology-derived parts, systems, and entire organisms. This approach can integrate different disciplines (biology, chemistry, physics, mathematics) and 'converging technologies' (biotechnology, nanotechnology, information technology); and knowledge derived from systems biology, whole-genome engineering, pathway engineering, mathematical modeling and computer-aided design, as well as the notion of interchangeable 'biological parts', are often seen as hallmarks of the SB idea (Lorenzo and Danchin 2008; NBT 2009; Way et al. 2014).

Considerable societal benefits from SB and its applications have been promised, including chemicals and fuels from renewable sources with lower greenhouse gas emissions, new therapies for diseases, and novel and rapidly deployable vaccines (Khalil and Collins 2010; Robertson et al. 2011; Ruder et al. 2011; Weber and Fussenegger 2012) – all of which could contribute to a new bioeconomy (OECD 2009; 2011). Conversely, critics have pointed out potential risks, be they biosafety and biosecurity risks (Church 2004; ETC 2007; Garfinkel et al. 2007; Kelle 2009; NRC 2006), environmental risks (e.g., negative impacts on biodiversity) or socioeconomic risks (e.g., food and water security, 'land grab') (Buyx and Tait 2011; ETC 2007; 2010; FoE 2010). Furthermore, ethical and other philosophical concerns about SB's effects on traditional notions of life are the subject of some debate (Boldt and Müller 2008; Cho et al. 1999).

Various recent studies have taken a closer look at the state of research and applications in SB (König et al. 2013; Rabinovitch-Deere et al. 2013; Way et al. 2014; Weber and Fussenegger 2012). Compared to 'traditional' genetic engineering – which may be described as mostly enhancing existing biological functions or transferring them between organisms, based on the modification or transfer of one or very few genes – new complex functions based on the combination and functional integration of many genes have been linked to SB work. In some cases these complex functions depended significantly on computational design and modeling. Furthermore, genome synthesis/assembly approaches (synthetic genomics) and genome editing techniques allowed to simultaneously alter genes on the scale of the whole genome, and thereby enabled new applications that were not possible by more 'traditional' genetic engineering (where typically only very few nucleotides or genes in an organism are altered by recombinant DNA technology). Thus, increasingly complex gene circuits have been generated to detect toxic metals (Prindle et al. 2012) or multiple changes in cancer cells (Xie et al. 2011); or computer-modeled, sophisticated non-natural metabolic pathways to produce chemicals and fuels have been constructed (Bond-Watts et al. 2011; Yim et al. 2011; Zhang et al. 2012). Gene synthesis and/or genome assembly have been used to reconstruct viruses, including the polio virus or the virus of the 1918 influenza pandemic (Wimmer et al. 2009); or to introduce genome-wide changes for designing vaccine candidates from the polio virus and influenza viruses (Mueller et al. 2010; Wimmer et al. 2009). Similarly, synthetic genomics techniques have been used to generate a first bacterial (*Mycoplasma*) cell controlled by a chemically synthesized genome upon transplantation into a related recipient cell (Gibson et al. 2010), and to make a first synthetic eukaryotic chromosome in yeast (Annaluru et al. 2014). Furthermore, extensive genome editing has recently allowed to alter and to expand the genetic code in bacteria (Mandell et al. 2015; Rovner et al. 2015).

Looking at the state of research and development, it appears that we are currently dealing with a wide spectrum of approaches pertaining to, and proposed for, the SB idea. These approaches can also be part of different application schemes, including the production of chemicals from plant feedstocks in closed systems by genetically modified microorganisms (GMMs); strategies that would require the release of GMMs (e.g., for bioremediation); or the use of genome synthesis to generate viral vaccines for use in humans. Nevertheless, many of these approaches and the organisms generated from them may not be clearly distinguishable from 'traditional' genetic engineering and molecular biology approaches and their products (see, e.g., König et al. 2013; Nielsen and Keasling 2011). This circumstance poses an additional issue for any proposed regulation (and indeed communication) on SB or 'synthetic' organisms. The actual benefits and risks of the SB applications currently envisaged thus appear to depend on issues and challenges on different levels. In the following we will outline these various dimensions of the benefits and risks with a view to their implications for governance and policy schemes needed to foster opportunities and to responsibly govern potential transformations linked to SB applications.

2 Dimensions of benefits and risks

2.1 General issues associated with applications of SB: socioeconomics

The benefits and risks of important potential applications of SB appear to depend on general aspects of application schemes rather than on issues directly relating to SB. For example, the negative impacts on biodiversity or water/food security as a result of large-scale planting of energy crops, intended to be converted into biofuels by genetically engineered/'synthetic' organisms, may be no different to those associated with any other energy-conversion technology using such biomass (such as Fischer-Tropsch synthesis, catalytic processes, or pyrolysis schemes (Regalbuto 2009)). However, although these factors are not qualitatively new, a large increase in the scope of growing energy crops linked to a simpler and economically more attractive conversion to biofuels by 'synthetic' organisms may make these issues more pressing. Similarly, broad patents and patent thickets – the number of which might increase as a result of generating more and more complex functions and systems based on multiple biological parts and devices (Kumar 2007; Rutz 2009) – already pose a challenge in the biopharmaceutical industry (Zimmeren et al. 2011). Whether, and to what extent, such socioeconomic questions linked to these more general aspects of SB application schemes could affect societies will depend on various parameters. For instance, important challenges associated with the benefits of biobased chemicals or 'synthetic' biofuels (including the previously mentioned socioeconomic issues in poor countries) may depend on feedstock, product, processing schemes and land use (see König et al. 2013, and references therein). Furthermore, the way in which patents for SB solutions are organized and applied may influence the extent to which poor countries in the Global South – which are likely to be the main areas for (plant-derived) biomass production (Berndes et al. 2003) – have access to biofuel feedstocks and technologies (Juma and Bell 2009).

2.2 Possible SB-specific issues: biosafety and biosecurity

In contrast to these more general aspects relating to certain application schemes, a number of benefits and risks may be more specifically associated with (and likely directly affected by) SB approaches. These benefits and risks may arise from developments that include newly designed 'biological parts'; the construction of new (complex) signaling and metabolic pathways and circuitries; the use of non-natural molecules; genome editing; and whole genome synthesis and assembly. These developments and their possible effects may be further potentiated by advances in a central SB concept, namely computationally-guided modeling and design.

All these developments will allow the degree to which biological systems could be modified to be considerably increased – up to 'completely synthetic' organisms in the future. This may result in new challenges regarding concepts and traditional notions of life (which we will not outline further here; see, e.g., Boldt and Müller 2008) as well as the risk assessment of such organisms. It may thus become more difficult to assess the risks of extensively genetically modified or (putative) entirely 'synthetic' future organisms by the current concepts that are based on comparisons involving recipient organisms and the origin of the transferred genes. This issue may become more significant as methodologies in SB progress further.

Large-scale custom DNA synthesis (and new genome assembly techniques) combined with knowledge from functional genomics might also facilitate the generation and the malicious use of (new) pathogens. Similarly, the construction of new metabolic pathways guided by computational design – as already demonstrated for the production of certain chemicals and biofuels (Bond-Watts et al. 2011; Yim et al. 2011) – might allow the generation of bacteria or fungi that produce novel toxins, which could be used as new pathogens. Finally, the potential weaponization of pathogens by enhancing their stability and ability to survive (often considered as an important condition for their effective use as/in weapons) is the subject of debate (UNICRI 2012). These developments have raised important concerns in connection with SB. Moreover, the possibility that synthesized 'bioparts' – and their (envisioned) straightforward combination into new biological functions via computational design – could make SB accessible to a broader spectrum of actors (i.e., beyond nation-states) has augmented such concerns (Tucker 2011; UNICRI 2012).

2.3 Further conditions and aspects of SB development

In addition to the afore-mentioned two main dimensions of potential benefits and risks associated with SB applications, any assumption regarding future benefits and risks needs to be seen in the light of the presumably low predictability of the exact nature of future application schemes and innovations from a still emerging field. Other important aspects to be addressed by any governance scheme concern the global impacts that SB will likely develop. On the one hand, these impacts may be driven by requirements for large-scale biomass production – a prerequisite for a new transforming bioeconomy – which would have to come largely from the Global South, at least if such requirements depend on plant feedstocks (Berndes et al. 2003). On the other hand, knowledge, expertise, and equipment in biosciences and biotechnology appear to be rapidly proliferating. According to some experts, the proliferation of biological and biotechnological knowledge and equipment has already occurred on a large scale, both geographically and within societies. Such proliferation also involves players not associated with academic or industrial institutions (Tucker 2011; UNICRI 2012).

3 Implications for governance and policies

In order to foster opportunities from the field and responsibly deal with its challenges, governance and policy agendas for SB should be able to address the different dimensions of benefits and risks outlined in the previous paragraphs.

3.1 Socioeconomics

Curbing negative socioeconomic consequences from issues that are not directly linked to the specifics of SB (but depend rather on general aspects of SB application schemes on a global scale) may require that products or applications be subjected to broadly applicable and effective environmental, socioeconomic, and ethical standards. To be effective, these standards have to deal with the general dimension of the issue (e.g., conditions for generating biomass) rather than with the exact, though less important, nature of the underlying technical approach. In this way, negative impacts from biobased chemicals or fuels could be mitigated by applying international sustainability and human rights standards to their production (Buyx and Tait 2011). Similarly, intellectual property issues could be addressed by international governmental organizations and industry within the framework of collaborative licensing models (Juma and Bell 2009; Zimmeren et al. 2011). Current intergovernmental institutions dealing with 'global commons', such as the United Nations' Framework Convention on Climate Change (FCCC) and the Convention on Biological Diversity (CBD), seem like prototypic examples for developing and implementing such standards. Conversely, these models appear to be unable to efficiently advance the protection of 'global commons' – and appropriate governance schemes may need to experiment with new ideas (see, e.g., Vasconcelos et al. 2013, and references therein; Sukhdev 2012).

3.2 Biosecurity

As regards the issues more specifically linked to SB, most experts see potential biosecurity threats arising in the near future from the misuse of SB primarily in connection with state-supported players (including terrorist groups) and, above all, nation-states (Garfinkel et al. 2007; Suk et al. 2011; Tucker 2011; UNICRI 2012). It is considered rather unlikely that non-expert players (i.e., players without a thorough knowledge of methods and laboratory/experimental experience with corresponding pathogens) could use SB to generate pathogens or toxins, multiply those pathogens to obtain sufficient amounts and keep them infectious for generating weapons (Jefferson et al. 2014; Suk et al. 2011; Tucker 2011; UNICRI 2012). Yet the differences between non-state or non-expert players and state players that underlie such assessments might become smaller or even vanish in the middle or long term, if (targeted) engineering of biological systems were to become increasingly simpler and less dependent on expert knowledge ('de-skilling') (Tucker 2011; UNICRI 2012). It is not clear, however, whether, and to what extent, such an evolution in SB technology could compensate/substitute laboratory experience and tacit knowledge (Jefferson et al. 2014; Tucker 2011).

The production of biological weapons using SB approaches would be forbidden by the Biological and Toxin Weapons Convention (BTWC) (http://www.opbw.org/; BTWC 1986) and, with regard to toxins, by the Chemical Weapons Convention (CWC, n.d.). However, no compliance/control mechanisms exist in the BTWC, the convention most relevant for SB (Tucker 2010). Thus, there are no international mechanisms to prevent the development and proliferation of technologies for the production of biological weapons, apart from the voluntary agreements of the (relatively small) Australia Group, an informal forum of countries that seeks to harmonize export controls so that exports do not contribute to the development of chemical or biological weapons (www.australiagroup.net). In addition, two international associations of gene synthesis companies have established codes of conduct to screen made-to-order genes for pathogen sequences and to verify the legitimacy of customers (IASB 2009; IGSC 2009). In some countries, exports of equipment, materials, technology, software, and data that could lead to the production of biological and chemical weapons are further subjected to specific regulations, such as EC Council Regulation No 428/2009 (EC 2009) and national export regulations based thereon in the European Union.

Control mechanisms similar to those associated with the CWC or nuclear non-proliferation, including a body comparable to the International Atomic Energy Agency (IAEA), have been proposed (Tucker 2010; UNICRI 2012). At the same time, the feasibility of efficient control is seen rather skeptically: in contrast to the development of chemical and nuclear weapons, the same materials and equipment would be expected to be used in both civil biological research and biotechnology linked to military or malicious use (Garfinkel et al. 2007; Tucker 2010; UNICRI 2012). Furthermore, a myriad of (known) laboratories would have to be controlled.

Finally, a major issue regarding biosecurity is the control of dual-use knowledge. Recent experiments, involving directed evolution and genetic engineering, on the airborne transmission of bird-flu viruses in ferrets (a model to study influenza transmission in humans) (Herfst et al. 2012; Imai et al. 2012; Linster et al. 2014; Watanabe et al. 2014) have reinvigorated this debate. Crucial points of concern in the control of dual-use knowledge include the role of potential benefits and harms from biosecurity-sensitive research results, and whether and how access to such data should, or could (given, for instance, the power of global online networks), be restricted or prevented. And, if deemed possible, who should decide on these points (see, e.g., Fauci and Collins 2012; Malakoff 2013). Policies may need to incorporate more effective rules by governments and funding agencies to identify critical experiments before they are funded; and thus before they are started. Efforts to introduce such rules have recently been undertaken in the USA (USGov 2014).

In conclusion, any program to prevent the misuse of biotechnology and SB is faced with considerable technical issues linked to the control of materials and equipment, but also with conceptional and ethical challenges associated with the potential control of dual-use knowledge. Hence, in addition to efforts to establish more effective international conventions on biological and chemical weapons as well as export rules on dual-use technologies (including data and knowledge relating to such technologies), further governance measures appear to be required. These should strive to create a culture of shared responsibility and awareness regarding biosecurity issues. Such measures may involve or address players from academic and non-academic research communities (including biohacker/DIY communities), policymakers, governmental agencies, technology assessment institutions, civil society organizations, and the public. This may require establishing education schemes and structures designed to promote open dialog, with avenues for mutual learning among internal and external players, including civil society/the public(s). The measures should empower as many (and diverse) players as possible to recognize potential misuse; and to assume responsibility for contributing to the development and implementation of solutions that can prevent or mitigate the malicious use of biotechnologies and SB approaches.

3.3 Biosafety

The expected increase in the degree to which biological systems can be modified – up to the envisioned 'completely synthetic' organisms – may appear the most obvious challenge relating to biosafety risk assessment. So far, this assessment has been based on the recipient organisms, gene transfer vectors and transferred genes (from donor organisms), as well as – in the case of deliberative release of genetically modified organisms (GMOs) – a comparative analysis involving non-GM organisms with a history of safe use (Pauwels et al. 2013). Such an assessment may become increasingly difficult, or even impossible, if organisms will be constructed based on newly designed or non-natural 'biological parts', and/or will become more and more different from existing (comparator) organisms by an increasing number of genetic modifications. The extent to which this issue will actually become relevant to SB approaches in the (near) future, and would make changes in the current regulations necessary, remains however unclear. This is mainly due to the emerging character of the field and the presumably low predictability of the exact nature of future innovations and progress in the various SB approaches.

Current approaches to biotechnological applications of SB are still based on well-known biotechnology 'work horses'. For instance, this is the case with strains of *E. coli* or baker's yeast, even when they contain complex, so-called 'synthetic' metabolic pathways (see, e.g., Bond-Watts et al. 2011; Yim et al. 2011; Zhang et al. 2012); have a reduced genome size (Leprince et al. 2012; Pósfai et al. 2006) in order to produce minimal cells as a 'chassis' for biotechnological producer strains; or are 'genetically recoded' as part of biocontainment strategies (Mandell et al. 2015; Rovner et al. 2015). Similarly, microalgae metabolically engineered to produce fuels are, so far, rather close to parental genera (see, e.g., Reppas and Ridley 2010; Roessler et al. 2009). As regards approaches to 'synthesize' primitive models for cells (protocells) 'from scratch' (using chemicals such as nucleotides or membrane components not used by extant cells), the current status shows that attempts to create self-reproducing entities have not led to any cell-like system capable of gene-dependent self-replication (Blain and Szostak 2014). Furthermore, semi-synthetic approaches to cellular systems are still based on purified components from existing cells, and so far only recapitulate few functions of known living systems (Blain and Szostak 2014; Stano and Luisi 2013).

Thus, current work linked to SB appears to pose no completely new biosafety challenges. In the European Union (EU), all current approaches – apart from still nascent work aiming at 'synthesizing' cells/organisms *in vitro* or 'from scratch' (i.e., bottom-up) – are covered by existing EU regulations (see also the first interim report on SB of Germany's Central Committee on Biological Safety (ZKBS 2012), and recent opinions of European biosafety advisory bodies and experts (Pauwels et al. 2013; SCHER 2014)). The regulations encompass several EU directives, notably the directives on the contained use of genetically modified micro-organisms (EC 1990; 1998) and on the deliberate release into the environment of GMOs (EC 2001; Buhk 2014). In the mid to long term, however, it may be necessary to adapt these regulations with regard to what is considered an organism and what is classed as a GMO, if work generating 'synthetic' and 'semi-synthetic' cells by bottom-up approaches should progress accordingly (see also Buhk 2014; Pauwels et al. 2013). Furthermore, there may be the need to develop risk assessment methods that could deal with the challenges that (future) GMOs – which should increasingly differ from existing organisms and, at the same time, might be produced in high numbers by advancing SB approaches – could pose (conceptionally and/or with respect to assessment capacities) to the current 'comparative' and 'case-by-case' risk assessment (Pauwels et al. 2013; SCHER 2014). Issues for risk assessment derived from GMOs with increasingly complex genetic modifications and from a potentially high number of GMOs for commercial use have also been identified as key challenges to the existing regulatory system in the USA (Carter et al. 2014).

But certain points may be worth considering even today. These may include more offers by governments or local authorities to biohacker/DIY communities for information/advice about biosafety and for public working spaces. Furthermore, criticism about insufficient conflict-of-interest management on the part of the European Court of Auditors (ECA 2012) – such as the occurrence of 'revolving door' situations in European agencies, including EFSA (which deals with GMOs) – suggests that there is room for improvement in the governance of these regulatory agencies. Finally, the long-term behavior of GMOs and their environmental effects could be hard to test and assess definitively, even if reliable biological containment/firewall mechanisms (Dolgin 2015; Wright et al. 2013) were to be developed in future. It might, therefore, be worth considering possibilities for tracking risks and assigning liabilities (Lloyd's 2009) in connection with damage caused by the release of such organisms. 'Gene watermarking' of synthetic DNA sequences in organisms (Gibson et al. 2010; Liss et al. 2012) could serve as a means to allow such tracking and help to identify manufacturers and users. Mandatory watermarking could, for instance, be considered for 'synthetic' organisms/GMOs used in applications with a high risk of such organisms escaping into the environment (e.g., algae in open ponds) or that require the environmental release of such organisms or their direct application in humans.

3.4 General policy aspects

In view of the multiple dimensions that can be linked to both opportunities and challenges arising from SB and its applications, effective policies should benefit from being informed by diverse perspectives and expertise. Knowledge from experts of different scientific disciplines as well as knowledge generated by dialog with a broad spectrum of stakeholders and the public could serve as valuable elements of such a policy-informing process. However, it would also be necessary to create the conditions and structures that can encourage and empower different players/stakeholders to contribute to such a process. Dialog initiatives such as the EU-funded SYNENERGENE project (www.synenergene.eu) may provide hints to developing such conditions. Furthermore, even if appropriate conditions for such a proposed multidisciplinary and transdisciplinary policy input could be obtained, this might not suffice to generate the desired efficient policy output.

As we have recently proposed, potential shortcomings in a state's political system (including governments, parliaments, and regulatory agencies) can be factors to make policy output inefficient (König et al. 2014). These may include phenomena such as regulatory capture, including 'revolving door' situations (Bó 2006; Shapiro 2012), but also special interests of states and governments in their 'own ventures' through investments or stakes in companies (Da Rin et al. 2011; *The Economist* 2012; Pargendler 2012). Conflicts of interest stemming from the role of state stakeholders as owners and regulators of their ventures might become especially relevant in the energy sector, which is said to harbor a huge potential for SB, and/or in emerging economies (OECD 2009, 2011; Wooldridge 2012). Moreover, national bioeconomical and military defense interests may be factors that interfere with efficient policy output. These have, for instance, been linked to the lack of compliance measures in the Biological Weapons Convention (Tucker 2010). Such issues deriving from the relationships between state stakeholders, vested interests from inside and outside the political system, and technology development may add a further dimension to the challenges that need to be taken into account in governance strategies for SB and other emerging technologies.

These diverse challenges – as well as the uncertainties and unknowns from a nascent technoscience interwoven with these challenges – lead us to suggest that politics-driven and programmatic (and purely application-directed) strategies to 'construct' specific research or innovations may be of limited use for governance

(Frank et al. 2015; König et al. 2014). Rather, we propose the need for an exploration-based concept towards 'cultures of responsible experimentation' (CORE). Corresponding innovation and safety cultures should strive to create conditions and stimuli that can encourage creativity and broad experimentation. Such experimentation would relate to science/technology, risk assessment and management, but also to corporate and political culture. As an element of an evolution-like process guided by a framework of overriding ethical values, such diverse exploration should enhance the probability of finding appropriate solutions to grand societal challenges (König et al. 2014, and references therein). Responsibility for the guiding ethical framework (i.e., for its constituents, responsiveness, or shaping impact) would require contributions from various players. Thus, bold and visionary political and business leaders could contribute by inspiring their peers to pursue experimental pathways. Ultimately, however, such pathways should produce economic and social benefits for a wide range of people in societies. These benefits could generate stimuli from civil society to governments and policy-making bodies, and facilitate the implementation of new pathways.

References

Annaluru, N., Muller, H., Mitchell, L.A., Ramalingam, S., Stracquadanio, G., Richardson, S.M.,... Chandrasegaran, S. (2014). Total Synthesis of a Functional Designer Eukaryotic Chromosome. *Science, 344*, 55-58. doi: 10.1126/science.1249252.
Berndes, G., Hoogwijk, M., & Broek, R. (2003). The contribution of biomass in the future global energy supply: a review of 17 studies. *Biomass and Bioenergy, 25*, 1-28.
Blain, J.C., & Szostak, J.W. (2014). Progress Toward Synthetic Cells. *Annual Review of Biochemistry, 83*, 615-640.
Bó, E.D. (2006). Regulatory capture: a review. *Oxford Review of Economic Policy, 22*, 203-225.
Boldt, J., & Müller, O. (2008). Newtons of the leaves of grass. *Nature Biotechnology, 26*, 387-389.
Bond-Watts, B.B., Bellerose, R.J., & Chang, M.C.Y. (2011). Enzyme mechanism as a kinetic control element for designing synthetic biofuel pathways. *Nature Chemical Biology, 7*, 222-227.
BTWC (1986). Second Review Conference Final Declaration. Biological and Toxin Weapons Convention. www.opbw.org/rev_cons/2rc/docs/final_dec/2RC_final_dec_E.pdf. Accessed: 26 March 2015.
Buhk, H.J. (2014). Synthetic biology and its regulation in the European Union. *New Biotechnology, 31*, 528-531.
Buyx, A., & Tait, J. (2011). Ethics. Ethical framework for biofuels. *Science, 332*, 540-541.
Carter, S.R., Rodemeyer, M., Garfinkel, M.S., & Friedman, R.M. (2014). Synthetic Biology and the US Biotechnology Regulatory System: Challenges and Options. J. Craig Venter

Institute. http://www.jcvi.org/cms/fileadmin/site/research/projects/synthetic-biology-and-the-us-regulatory-system/full-report.pdf. Accessed: 26 March 2015.
Cho, M.K., Magnus, D., Caplan, A.L., & McGee, D. (1999). Policy forum: genetics. Ethical considerations in synthesizing a minimal genome. *Science, 286,* 2089-2090.
Church, G. (2004). A Synthetic Biohazard Non-proliferation Proposal. http://arep.med.harvard.edu/SBP/Church_Biohazard04c.htm. Accessed: 26 March 2015.
CWC (n.d.). Chemical Weapons Convention. Organisation for the Prohibition of Chemical Weapons. https://www.opcw.org/chemical-weapons-convention/. Accessed: 26 March 2015.
Da Rin, M., Hellmann, T.F., & Puri, M. (2011). A survey of venture capital research. National Bureau of Economic Research. Working Paper 17523. http://www.nber.org/papers/w17523. Accessed: 26 March 2015.
Dolgin, Elie. (2015). GM microbes created that can't escape the lab. *Nature, 517,* 423.
EC (1990). European Council. Council Directive of 23 April 1990 on the contained use of genetically modified micro-organisms (90/219/EEC), OJ 1990 L 117/1.
EC (1998). European Council. Council Directive 98/81/EC of 26 October 1998 amending Directive 90/219/EEC on the contained use of genetically modified micro-organisms, OJ 1998 L 330/13.
EC (2001). European Council. Directive 2001/18/EC of the European Parliament and of the Council of 12 March 2001 on the deliberate release into the environment of genetically modified organisms and repealing Council directive 90/220/EEC, OJ 2001 L 106/1.
EC (2009). European Council. Council Regulation (EC) No 428/2009 of 5 May 2009 setting up a Community regime for the control of exports, transfer, brokering and transit of dual-use items, OJ 2009 L 134/1.
ECA (2012). Management of conflict of interest in selected EU Agencies. European Court of Auditors. Special report No 15/2012. doi: 10.2865/21104. http://bookshop.europa.eu/en/management-of-conflict-of-interest-in-selected-eu-agencies-pbQJAB12014/. Accessed: 26 March 2015.
The Economist (2012). European venture capital. Venturecrats. *The Economist,* 19 April 2012. http://www.economist.com/blogs/schumpeter/2012/04/european-venture-capital. Accessed: 26 March 2015.
ETC (2007). ETC Group. *Extreme Genetic Engineering. An Introduction to Synthetic Biology.* http://www.etcgroup.org/sites/www.etcgroup.org/files/publication/602/01/synbioreportweb.pdf. Accessed: 26 March 2015.
ETC (2010). ETC Group. *The New Biomassters. Synthetic Biology and the Next Assault on Biodiversity and Livelihoods.* http://www.etcgroup.org/sites/www.etcgroup.org/files/biomassters_27feb2011.pdf. Accessed: 26 March 2015.
Fauci, A. S., & Collins, F. S. (2012). Benefits and risks of influenza research: lessons learned. *Science, 336,* 1522-1523.
FoE (2010). Friends of the Earth. *Synthetic Solutions to the Climate Crisis: The Dangers of Synthetic Biology for Biofuels Production.* http://libcloud.s3.amazonaws.com/93/59/9/529/1/SynBio-Biofuels_Report_Web.pdf. Accessed: 26 March 2015.
Frank, D., Heil, R., Coenen, C., & König, H. (2015). Synthetic biology's self-fulfilling prophecy - dangers of confinement from within and outside. *Biotechnology Journal,* 10, 231-235.
Garfinkel, M.S., Endy, D., Epstein, G.L., & Friedman, R.M. (2007). Synthetic genomics | options for governance. *Biosecurity Bioterrorism,* 5, 359-362.

Gibson, D.G., Glass, J.I., Lartigue, C., Noskov, V.N., Chuang, R.Y., Algire, M.A., ... Venter, J.C. (2010). Creation of a bacterial cell controlled by a chemically synthesized genome. *Science, 329*(5987), 52-56. doi: 10.1126/science.1190719.
Herfst, S., Schrauwen, E.J., Linster, M., Chutinimitkul, S., Wit, E. d., Munster, V.J., ... Fouchier, R.A. (2012). Airborne transmission of influenza A/H5N1 virus between ferrets. *Science, 336*(6088), 1534-1541.
IASB (2009). International Association Synthetic Biology. *The IASB Code of Conduct for Best Practices in Gene Synthesis*. www.ia-sb.eu/tasks/sites/synthetic-biology/assets/File/pdf/iasb_code_of_conduct_final.pdf. Accessed: 26 March 2015.
IGSC (2009). International Gene Synthesis Consortium. *Harmonized screening protocol: gene sequence & customer screening to promote biosecurity*. http://www.genesynthesisconsortium.org/images/pdf/IGSC%20Harmonized%20Screening%20Protocol-11_18_09.pdf. Accessed: 26 March 2015.
Imai, M., Watanabe, T., Hatta, M., Das, S.C., Ozawa, M., Shinya, K., ... Kawaoka, Y. (2012). Experimental adaptation of an influenza H5 HA confers respiratory droplet transmission to a reassortant H5 HA/H1N1 virus in ferrets. *Nature, 486*(7403), 420-428.
Jefferson, C., Lentzos, F., & Marris, C. (2014). Synthetic biology and biosecurity: challenging the "myths". *Frontiers in Public Health, 2*, 115. doi: 10.3389/fpubh.2014.00115.
Juma, C., & Bell, B. (2009). V. Advanced Biofuels and Developing Countries: Intellectual Property Scenarios and Policy Implications. In: *The Biofuels Market: Current Situation and Alternative Scenarios* (pp.63-89). United Nations report UNCTAD/DITC/BCC/2009/1. New York: United Nations.
Kelle, A. (2009). Ensuring the security of synthetic biology-towards a 5P governance strategy. *Systems and Synthetic Biology, 3*, 85-90.
Khalil, A.S., & Collins, J.J. (2010). Synthetic biology: applications come of age. *Nature Reviews Genetics, 11*, 367-379.
König, H., Frank, D., & Heil, R. (2014). Science, technology and the state: implications for governance of synthetic biology and emerging technologies. In: T. Michalek, L. Hebáková, L. Hennen, C. Scherz, L. Nierling & J. Hahn (eds.), *Technology Assessment and Policy Areas of Great Transitions. Proceedings from the PACITA 2013 Conference in Prague*. Prague: Technology Centre ASCR.
König, H., Frank, D., Heil, R., & Coenen, C. (2013). Synthetic Genomics and Synthetic Biology Applications Between Hopes and Concerns. *Current Genomics, 14*, 11-24.
Kumar, S. (2007). Synthetic Biology: The Intellectual Property Puzzle. *Texas Law Review, 85*, 1745-1768.
Leprince, A., Lorenzo, V. d., Voller, P., Passel, M.W. v., & Martins dos Santos, V.A. (2012). Random and cyclical deletion of large DNA segments in the genome of Pseudomonas putida. *Environmental Microbiology, 14*, 1444-1453.
Linster, M., Boheemen, S. v., Graaf, M. d., Schrauwen, E.J., Lexmond, P., Mänz, B., ... Herfst, S. (2014). Identification, characterization, and natural selection of mutations driving airborne transmission of A/H5N1 virus. *Cell, 157*(2), 329-339.
Liss, M., Daubert, D., Brunner, K., Kliche, K., Hammes, U., Leiherer, A., & Wagner, R. (2012). Embedding permanent watermarks in synthetic genes. *PLOS ONE, 7*, e42465. doi: 10.1371/journal.pone.0042465.
Lloyd's (2009). *Synthetic biology: influencing development. Lloyd's emerging risks team report*. http://www.lloyds.com/~/media/lloyds/reports/emerging%20risk%20reports/syntheticbiology_influencethedebate_july2009_v1.pdf. Accessed: 26 March 2015.

Lorenzo, V. d., & Danchin, A. (2008). Synthetic biology: discovering new worlds and new words. *EMBO Reports, 9*, 822-827.

Malakoff, D. (2013). Avian influenza. Critics skeptical as flu scientists argue for controversial H7N9 studies. *Science, 341*, 601.

Mandell, D.J., Lajoie, M.J., Mee, M.T., Takeuchi, R., Kuznetsov, G., Norville, J.E., Gregg, C.J., Stoddard, B.L., & Church, G.M. (2015). Biocontainment of genetically modified organisms by synthetic protein design. *Nature, 518*, 55-60.

Mueller, S., Coleman, J.R., Papamichail, D., Ward, C. B., Nimnual, A., Futcher, B., Skiena S., & Wimmer, E. (2010). Live attenuated influenza virus vaccines by computer-aided rational design. *Nature Biotechnology, 28*, 723-726.

NBT (2009). What's in a name? *Nature Biotechnology, 27*, 1071-1073.

Nielsen, J., & Keasling, J. D. (2011). Synergies between synthetic biology and metabolic engineering. *Nature Biotechnology, 29*, 693-695.

NRC (2006). National Research Council. *Globalization, Biosecurity, and the Future of the Life Sciences.* Washington, D.C.: The National Academies Press.

OECD (2009). *The Bioeconomy to 2030: Designing a Policy Agenda.* Paris: OECD Publishing. doi: 10.1787/9789264056886-en.

OECD (2011). *Future Prospects for Industrial Biotechnology.* Paris: OECD Publishing. doi: 10.1787/9789264126633-en.

Pargendler, M. (2012). State Ownership and Corporate Governance. *Fordham Law Review, 80*, 2917-2973.

Pauwels, K., Mampuys, R., Golstein, C., Breyer, D., Herman, P., Kaspari, M., Pagès, J.-C., Pfister, H., Wilk, F. v. d., & Schönig, B. (2013). Event report: SynBio Workshop (Paris 2012)–Risk assessment challenges of Synthetic Biology. *Journal für Verbraucherschutz und Lebensmittelsicherheit, 8*, 215-226.

Pósfai, G., Plunkett, G., Fehér, T., Frisch, D., Keil, G.M., Umenhoffer, K., ... Blattner, F.R. (2006). Emergent properties of reduced-genome Escherichia coli. *Science, 312*(5776), 1044-1046.

Prindle, A., Samayoa, P., Razinkov, I., Danino, T., Tsimring, L. S., & Hasty, J. (2012). A sensing array of radically coupled genetic 'biopixels'. *Nature, 481*, 39-44.

Rabinovitch-Deere, C.A., Oliver, J.W., Rodriguez, G.M., & Atsumi, S. (2013). Synthetic biology and metabolic engineering approaches to produce biofuels. *Chemical Reviews, 113*, 4611-4632.

Regalbuto, J.R. (2009). Engineering. Cellulosic biofuels—got gasoline? *Science, 325*, 822-824.

Reppas, N.B., & Ridley, C.P. (2010). Methods and compositions for the recombinant biosynthesis of n-alkanes. *U.S. Patent No. 7,794,969.* Washington, D.C.: U.S. Patent and Trademark Office.

Robertson, D.E., Jacobson, S.A., Morgan, F., Berry, D., Church, G.M., & Afeyan, N.B. (2011). A new dawn for industrial photosynthesis. *Photosynthesis Research, 107*, 269-277.

Roessler, P.G., Chen, Y., Liu, B., & Dodge, C.N. (2009). Secretion of fatty acids by photosynthetic microorganisms. *United States Patent Application 20090298143.*

Rovner, A.J., Haimovich, A.D., Katz, S.R., Li, Z., Grome, M.W., Gassaway, B.M., ... Isaacs, F.J. (2015). Recoded organisms engineered to depend on synthetic amino acids. *Nature, 518*, 89-93.

Ruder, W.C., Lu, T., & Collins, J.J. (2011). Synthetic biology moving into the clinic. *Science, 333*, 1248-1252.

Rutz, B. (2009). Synthetic biology and patents. A European perspective. *EMBO Reports, 10*(1S), 14-17.
SCHER (2014). Scientific Committee on Health and Environmental Risks (SCHER), Scientific Committee on Emerging and Newly Identified Health Risks (SCENIHR), Scientific Committee on Consumer Safety (SCCS). *Preliminary Opinion on Synthetic Biology II: Risk assessment methodologies and safety aspects.* European Union. doi: 10.2772/63529.
Shapiro, S. (2012). The Complexity of Regulatory Capture: Diagnosis, Causality and Remediation. *Roger Williams University Law Review, 102*, 101-137.
Stano, P., & Luisi, P.L. (2013). Semi-synthetic minimal cells: origin and recent developments. *Current Opinion in Biotechnology, 24*, 633-638.
Suk, J.E., Zmorzynska, A., Hunger, I., Biederbick, W., Sasse, J., Maidhof, H., & Semenza, J. C. (2011). Dual-use research and technological diffusion: reconsidering the bioterrorism threat spectrum. *PLoS Pathogens, 7*, e1001253. doi: 10.1371/journal.ppat.1001253.
Sukhdev, P. (2012) Sustainability: The corporate climate overhaul. *Nature, 486*, 27-28.
Tucker, J.B. (2010). Seeking Biosecurity Without Verification: The New U.S. Strategy on Biothreats. *Arms Control Today (January/February 2010)*, 8-14.
Tucker, J.B. (2011). Could Terrorists Exploit Synthetic Biology? *The New Atlantis*, 69-81.
UNICRI (2012). United Nations Interregional Crime and Justice Research Institute. *Security Implications of Synthetic Biology and Nanobiotechnology. A Risk and Response Assessment of Advances in Biotechnology.* Turin: UNICRI. http://www.unicri.it/in_focus/files/UNICRI%202012%20Security%20Implications%20of%20Synthetic%20Biology%20and%20Nanobiotechnology%20Final%20Public-1.pdf. Accessed: 26 March 2015.
USGov (2014). *United States Government Policy for Institutional Oversight of Life Sciences Dual Use Research of Concern.* http://www.phe.gov/s3/dualuse/Documents/durc-policy.pdf. Accessed: 26 March 2015.
Vasconcelos, V.V., Santos, F.C., & Pacheco, J.M. (2013) A bottom-up institutional approach to cooperative governance of risky commons. *Nature Climate Change, 3*, 797-801.
Watanabe, T., Zhong, G., Russell, C.A., Nakajima, N., Hatta, M., Hanson, A., . . . Kawaoka, Y. (2014). Circulating avian influenza viruses closely related to the 1918 virus have pandemic potential. *Cell host & microbe, 15*(6), 692-705.
Way, J.C., Collins, J.J., Keasling, J.D., & Silver, P.A. (2014). Integrating biological redesign: where synthetic biology came from and where it needs to go. *Cell, 157*, 151-161.
Weber, W., & Fussenegger, M. (2012). Emerging biomedical applications of synthetic biology. *Nature Reviews Genetics, 13*, 21-35.
Wimmer, E., Mueller, S., Tumpey, T.M., & Taubenberger, J.K. (2009). Synthetic viruses: a new opportunity to understand and prevent viral disease. *Nature Biotechnology, 27*, 1163-1172.
Wooldridge, A. (2012). State capitalism. Special report. *The Economist*, 21 January 2012.
Wright, O., Stan, G.B., & Ellis, T. (2013). Building-in biosafety for synthetic biology. *Microbiology, 159*, 1221-1235.
Xie, Z., Wroblewska, L., Prochazka, L., Weiss, R., & Benenson, Y. (2011). Multi-input RNAi-based logic circuit for identification of specific cancer cells. *Science, 333*, 1307-1311.
Yim, H., Haselbeck, R., Niu, W., Pujol-Baxley, C., Burgard, A., Boldt, J., . . . Dien, S. v. (2011). Metabolic engineering of Escherichia coli for direct production of 1, 4-butanediol. *Nature chemical biology, 7*(7), 445-452.
Zhang, F., Carothers, J.M., & Keasling, J.D. (2012). Design of a dynamic sensor-regulator system for production of chemicals and fuels derived from fatty acids. *Nature Biotechnology, 30*, 354-359.

Zimmeren, E. v., Vanneste, S., Matthijs, G., Vanhaverbeke, W., & Overwalle, G. v. (2011). Patent pools and clearinghouses in the life sciences. *Trends in Biotechnology, 29,* 569-576.

ZKBS (2012). Zentrale Kommision für die Biologische Sicherheit (Central Commission for Biological Safety and Security). *Monitoring der Synthetischen Biologie in Deutschland. 1. Zwischenbericht der Zentralen Kommission für die Biologische Sicherheit vom 6. November 2012.* http://www.bvl.bund.de/SharedDocs/Downloads/06_Gentechnik/ZKBS/01_Allgemeine_Stellungnahmen_deutsch/01_allgemeine_Themen/Synthetische_Biologie.pdf?__blob=publicationFile&v=3. Accessed: 26 March 2015.

Contributors

Johannes Achatz • Centre for Ethics Jena, Friedrich Schiller University Jena, Jena, Germany

Bernadette Bensaude Vincent • Cetcopra, Université Paris 1 Panthéon-Sorbonne, Paris, France

Joachim Boldt • Department of Medical Ethics and the History of Medicine, Freiburg University, Freiburg im Breisgau, Germany

Matthias Braun • Department of Theology, Friedrich-Alexander-University Erlangen-Nürnberg, Erlangen, Germany

Peter Dabrock • Department of Theology, Friedrich-Alexander-University Erlangen-Nürnberg, Erlangen, Germany

Tobias Eichinger • Institute of Biomedical Ethics and History of Medicine, University of Zürich, Zürich, Switzerland

Bernd Giese • Department of Technology Development and Design, Faculty of Production Engineering, University of Bremen, Bremen, Germany

Arnim von Gleich • Department of Technology Development and Design, Faculty of Production Engineering, University of Bremen, Bremen, Germany

Harald König • Institute for Technology Assessment and Systems Analysis (ITAS), Karlsruhe Institute of Technology (KIT), Karlsruhe, Germany

Sacha Loeve • TSH-COSTECH, Compiègne Technology University, Compiègne, France

Harald Matern • Faculty of Theology, University of Basel, Basel, Switzerland

Iñigo de Miguel Beriain • Inter-University Chair in Law and the Human Genome, University of Deusto, University of the Basque Country, Bilbao, Spain

Oliver Müller • Department of Philosophy, University of Freiburg, Freiburg im Breisgau, Germany

Christian Pade • Department of Technology Development and Design, Faculty of Production Engineering, University of Bremen, Bremen, Germany

Rainer Paslack • SOKO Institute for Social Research and Communication, Bielefeld, Germany

Jens Ried • Department of Theology, Friedrich-Alexander-University Erlangen-Nürnberg, Erlangen, Germany

Jürgen Robienski • Center for Ethics and Law in the Life Sciences (CELLS), Leibniz University Hannover and Medical School Hannover, Hannover, Germany

Jan C. Schmidt • Department of Social Sciences, Darmstadt University of Applied Sciences, Darmstadt, Germany

Jürgen Simon • Center for Ethics and Law in the Life Sciences (CELLS), Leibniz University Hannover and Medical School Hannover, Hannover, Germany

Christoph Then • Testbiotech e.V., München, Germany

Henning Wigger • Department of Technology Development and Design, Faculty of Production Engineering, University of Bremen, Bremen, Germany

MIX
Papier aus verantwortungsvollen Quellen
Paper from responsible sources
FSC® C105338

If you have any concerns about our products,
you can contact us on
ProductSafety@springernature.com

In case Publisher is established outside the EU,
the EU authorized representative is:
**Springer Nature Customer Service Center GmbH
Europaplatz 3, 69115 Heidelberg, Germany**

Printed by Libri Plureos GmbH
in Hamburg, Germany